The
Origins
of
Natural Science
in
America

The
Origins
of
Natural Science
in
America

The Essays of George Brown Goode

Edited and with an Introduction by

Sally Gregory Kohlstedt

SMITHSONIAN INSTITUTION PRESS

Washington and London

© 1991 by Smithsonian Institution
All rights reserved
Edited by J. Thomas Dutro, Jr., and Nancy P. Dutro
Designed by Janice Wheeler

Library of Congress Cataloging-in-Publication Data
Goode, G. Brown (George Brown), 1851–1896.
    The origins of natural science in America : essays of George Brown
Goode / edited and with an introduction by Sally Gregory Kohlstedt.
        p.   cm.
Essays originally published as a collection in 1901.
Includes bibliographical references and index.
ISBN 1-56098-098-2
    1. Science museums—United States—History.   2. Science museums—United
States—Educational aspects.   3. Science museums—Philosophy.   4. Goode, G.
Brown (George Brown), 1851–1896.   5. Museum curators—United States—Biog-
raphy.   I. Kohlstedt, Sally Gregory, 1943–   .   II. Title.
Q105.U5G66   1991
507.4'73—dc20   91-10266

British Library Cataloging-in-Publication Data available
Manufactured in the United States of America
99   98   97   96   95   93   92   91   5   4   3   2   1
For permission to reproduce any illustration appearing in this book, please correspond
with Smithsonian Institution Archives. The Smithsonian Institution Press does not
retain reproduction rights for these illustrations or maintain a file of addresses for
sources.
The paper used in this publication meets the minimum requirements of the American
National Standard for Permanence of Paper for Printed Library Materials Z39.48-1984.

CONTENTS

LIST OF ILLUSTRATIONS

The photographs and sketches of individuals are part of a project of George Brown Goode to document and illustrate the history of American science. Most of the originals are in several scrapbooks in his manuscripts in the Smithsonian Institution as well as in the memorial volume from which the essays in this volume have been reprinted.

George Brown Goode

Louis Agassiz

John James Audubon

Alexander Dallas Bache

Spencer Fullerton Baird

William Bartram

James Dwight Dana

John William Draper

Amos Eaton

Benjamin Franklin

Asa Gray

Joseph Henry

Edward Hitchcock

Benjamin Henry Latrobe

William Maclure

Ormsby MacKnight Mitchel

Samuel Latham Mitchill

Constantine Rafinesque

Charles V. Riley

David Rittenhouse

William Barton Rogers

Thomas Say

Henry Rowe Schoolcraft

Benjamin Silliman, Sr.

Sir Hans Sloane

John Torrey

Gerard Troost

Alexander Wilson

Jeffries Wyman

Edward Livingston Youmans

PREFACE

The essays reprinted in this volume provide renewed access to the ideas of George Brown Goode, the curator, historian, and museum administrator whose efforts shaped the United States National Museum in Washington, D.C. His practical example as museum curator and director within the Smithsonian Institution, coupled with his capacity to think very broadly and theoretically about museum principles and practice, established him as preeminent among his peers during the exceptionally active years of museum building and expansion in the last quarter of the nineteenth century.

Written a century ago, these essays of George Brown Goode demonstrate the outlook of a late Victorian struggling to understand the direction of contemporary science and to shape the future direction of scientific activity, using the educational and research institution he knew best. As a naturalist trained in the museum tradition, Goode believed that understanding historical processes was essential to describing the present state of scientific knowledge. He decided that the text of science should be presented in the context of the history of science. As Assistant Director and later Director of the National Museum established after the 1876 Centennial Exhibition, Goode implemented appropriate displays in what has today become the Arts and Industries Building, but which was originally built in the 1880s to house the Smithsonian's natural history and ethnological collections.

With virtually no historical studies that focussed on American science available to guide him, Goode set himself the task of compiling a history of natural history. What he discovered made him more conscious about the overall organization of the museum under his administration and led him to ponder as well the ways in which museums of the future might be structured. While his effort is necessarily quite preliminary, Goode's evidence and conclusions hold much that remains of value to historians and also to museum practitioners who plan for the museums of the twenty-first century.

The five essays reprinted here were originally presented as addresses before various scientific and historical audiences. Each is important for its contribution to history or to museum studies. Together, they reveal Goode's fundamental philosophy about the nature of history and the role of public museums. The first three offer a comprehensive overview of the development of the natural sciences, and to a lesser extent the physical sciences, from colonial times to the late nineteenth century. They conclude with Goode's rather pessimistic perspective on the changes taking place in his lifetime and reflect his distress about the way in which the comprehensive and collaborative naturalist tradition was being displaced by an emphasis on specialization and independent laboratory research. The latter two essays on museums are more optimistic and editorial, describing in somewhat linear fashion the progress of museum development and Goode's vision of what the future might hold if museums were able to function fully as both research and educational institutions.

These essays appear as they were originally reprinted for *A Memorial*

Volume of George Brown Goode, together with a Selection of His Papers on Museums and on the History of Science in America. This volume was published as part of the *Annual Report of the Board of Regents of the Smithsonian Institution, for the year ending June 30, 1897,* specifically as part two of the *Report of the United States National Museum* (Washington: Government Printing Office, 1901). The original memorial volume includes several eulogies by such eminent colleagues as Henry Fairfield Osborne, Samuel Pierpont Langley, and William Healy Dall; it also reprints other essays by Goode, including his "Principles of Museum Administration," and an extensive bibliography of his writings. It should be noted that the essays are reprinted here as they appeared, with no corrections or additions, except that the original footnotes have been renumbered and now appear as endnotes and a much needed index has been added.

GEORGE BROWN GOODE, 1851–1896

Sally Gregory Kohlstedt

George Brown Goode was internationally known among contemporaries for his role in museum development at the Smithsonian Institution in the late nineteenth century. His theories about the principles and practices of museum administration continue to be acknowledged, quoted, and debated. His research and institutional emphasis on the history of American science has been the starting point for later historians, particularly those who used his several pioneering essays on the early study of natural history. Despite his importance, however, no biography has yet been written to document his life and work. Even in histories of the Smithsonian Institution, Goode's activities have been overshadowed by fellow administrators who became Secretaries of the Institution and whose personalities—like the prominent physicist Joseph Henry, the gregarious biologist Spencer F. Baird, and the autocratic astronomer and physicist Samuel Pierpont Langley—are better remembered. Goode had an unusual capacity for leadership which, according to Langley, drew staff members and colleagues into cooperation with his principles and policies in order to establish the Smithsonian as a national cultural enterprise. His own quiet qualities of inspiration and reflection characterize the essays reprinted in this volume.[1]

I

Brown Goode, as he was called by his friends, was born in New Albany, Indiana, in 1851. Perhaps because he was an only child, he enjoyed private tutors and was the focus of parental attention as the family moved to Ohio, then to New York, and, during Goode's college years, to Connecticut. He graduated from Wesleyan University in 1870 and subsequently studied with the famous zoologist Louis Agassiz at the Museum of Comparative Zoology, a natural history research institution affiliated with Harvard College in Cambridge. Agassiz was a museum promoter without peer, and his infectious enthusiasm brought legislative and philanthropic support for the MCZ in Massachusetts. As the preeminent zoologist crossed the country on lecture tours, Agassiz helped local naturalists promote public and collegiate museums from Charleston to Chicago in the third quarter of the century. During this "harvest time of museums,"[2] federal expeditions also returned specimens to Washington that were widely distributed under Spencer Baird and helped invigorate regional and state enthusiasm for natural science. Equally important, the Cambridge museum became a training ground for many of the leading museum directors in the last half of the century, including Alpheus Hyatt of the Boston Museum of Natural History, Albert Bickmore at the American Museum of Natural History, Frederick Ward Putnam of the Peabody Museum at Harvard, and at least a half dozen other individuals at smaller museums and natural history societies. For museum staff, apprenticeship with an eminent naturalist remained the typical path to career advancement, a path quite distinct from peers aiming for an academic career who were going to Europe for advanced degrees in the last half of the century.[3]

Goode's first permanent position was that of curator of natural history specimens in the Orange Judd Hall of Natural Science at his alma mater. Sponsored by and dedicated to the successful agricultural editor who contributed a hundred thousand dollars to its construction, the new science building with its natural history museum was also intended to serve the community of Middletown. Throughout the 1870s, Goode worked to organize Wesleyan's college museum, adding to the collections through such standard techniques as small expeditions, exchanges with similar institutions, donations from loyal alumni and neighbors, and occasional purchases. Early annual reports of the museum contain the usual catalogue of acquisitions and activities, but they also reflect the young curator's self-conscious concern about the intended functions of a collegiate museum. Combining research and more general education seemed essential, and Goode portrayed the collections as a "comprehensive library of reference" that could service faculty, students, and the community. During his years at Wesleyan he married Judd's daughter, Sarah, herself a keen naturalist. By all accounts he was a serious young man, intent on his career and conscientious about public service.[4]

Tradition has it that the diffident youth was introduced to Spencer F. Baird through the initiative of his doting mother. However the meeting occurred, during the summer of 1872 Goode became a confidant of the older zoologist while working as an unpaid assistant on the United States Fish Commission expedition near Eastport, Maine. Subsequent summers were spent in the Bay of Fundy, Casco Bay, Noank on Long Island Sound, and elsewhere along the Atlantic coastline. Like other volunteers associated with Baird's enterprises, after these field expeditions Goode spent weeks doing research and writing up results in the various tower rooms of the Smithsonian "castle." Baird typically allowed his assistants to transfer selected duplicate specimens to their home institutions, in Goode's case to Wesleyan, for the use of college students and for public display.[5]

In 1875, Baird asked his ambitious assistant to take on the enormous responsibility for planning and organizing the Smithsonian's comprehensive zoological displays for the grand Centennial Exposition in Philadelphia. The project was unprecedented and far beyond Goode's earlier experience with international Fish Commission exhibitions. Baird hoped that a highly successful exhibit by the Smithsonian at the Philadelphia Exposition in 1876 would persuade Congressmen who attended that there should be a similar and permanent display in Washington. Moreover, the opportunity to acquire thousands of natural history specimens from abroad, donated by visiting nations, might provide additional leverage to pry money out of Congress. Baird's tactic was successful. A new museum building was authorized by Congress in 1878, and the Washington architectural firm of Claus and Schultz designed an ornate red brick building to stand next to the red sandstone castle on the Mall.[6] Goode became the Assistant Director, a position he held until he became Assistant Secretary of the Smithsonian Institution in charge of the museum, in 1887.

Until then, Baird retained the role of Director of the Museum and became, after the death of Joseph Henry in 1878, Secretary of the Smithsonian as well. In the meantime, Goode had taken over much of the day-to-day planning for the new museum, arranged for the transfer of specimens from Philadelphia, and initiated plans for the display and storage of the Smithsonian's holdings.[7] Following the example of his mentor and supervisor, he kept several projects running simultaneously. In 1880, he headed the United States Commission to the International Fisheries Exposition in Berlin. The trip had a secondary purpose as well. Baird put off finalizing plans for the interior of the new museum and urged Goode to visit European and British museums while abroad so that the United

States National Museum would incorporate the current museum features.[8] When the Assistant Director returned, he oversaw decisions
about everything from the design of display cases to the allocation of
space for each subject area for the new museum.[9]

Goode was a typical Victorian. The diffuse, eclectic enthusiasms which
resulted in asymmetrical castles were also reflected in multiple career
and avocational interests among intellectuals. Goode, no exception, can
appropriately be described as an ichthyologist, genealogist, historian, curator, and museum administrator. He published over two hundred articles, monographs, books, and annual reports during his relatively short
professional life. He proudly called himself a naturalist, publishing mostly
on fishes but also working on snakes and other animals. According to his
children, he was capable of identifying most living creatures found near
their home or during their vacations.

He insisted on thoroughness and efficiency in every aspect of his work.
Contemporaries were astonished by the effectiveness of his organizational
skills. As the Smithsonian became responsible for numerous American
exhibits at national and international fairs and exhibitions, Goode ration-
alized the process. He developed a system for packing materials in crates
that could be transformed into display tables, and within a couple of days
his staff could set up a polished public display with specimens and labels
in systematic order. These administrative and organizational approaches
served Goode well.[10]

His multifaceted career as a naturalist, museum administrator, and
historian was more exceptional in scale than in kind in late nineteenth-
century Washington, but his many interests kept Goode in the vortex of
intellectual and social life in the nation's capital. His memberships were
numerous and, according to colleagues, in many societies he was an active member, made scholarly presentations, and held various offices.
Thus, for example, he served as president of the Cosmos Club in 1893,
having been an active contributor to several committees and a leading

activist among those seeking to make the club facilities accessible to other local and national groups.[11] To do this, he worked long hours and on weekends, and was regularly chided by colleagues who warned him not to be overwhelmed by the "ceaseless grind of museum work."[12]

This extraordinarily busy life eventually took its toll. Throughout his adulthood, Goode had intermittent bouts of ill health which were often attributed to a nervous temperament and exhaustion. Sometimes, the symptoms were those of lethargy and depression, often called neurasthenia, and found among other young intellectuals like William James. Goode was also a heavy smoker. He died when he was forty-six years old from what contemporaries called overwork. His illness was identified as bronchitis by physicians but it may well have been emphysema.

Goode was a private man whose professional correspondence was cordial but restrained. Substantial official manuscript files in the Smithsonian archives reveal little personal correspondence with friends, and Goode apparently left no diary or journal that might document his personality and his interpersonal relationships. Perhaps his work was, as one eulogist put it, his life.[13]

Therefore, it is not surprising that his historical writing focused on the history of science in America and, to a considerable extent as well, on the history of the Smithsonian. Goode's historical outlook was influenced by his experience as a naturalist, by general and academic interest in history prevalent in the late nineteenth century, and by the practical requirements of being a museum curator and director.

II

Probably the first and fundamental influence on Goode's historical work was his research as a naturalist. Like other taxonomists who determine identity of and variation within and among species, he relied essentially on historical identifications and references in order to evaluate whether and by whom previous identifications of a particular specimen might have had been made. There was, as Agassiz had argued, "a world of meaning hidden under our zoological and botanical nomenclature," a historical legacy reinforced by crediting predecessors in reference notes and commemorating them in the naming of new species.[14]

Goode's earliest publications were in the normal practice of contemporary zoology. He concentrated on systematic descriptions of specimens acquired during his summer research efforts, particularly concentrating on the structure, life habits, and distributions of fishes. His catalogue of the fishes of Bermuda (made during a winter "recovery period" on that island) increased the number of recorded species from seven to seventy-five, and he provided a careful analysis of their probable geographical derivation. Biological problems such as migration, coloring, mimicry, and the relation of changes in environment to the population of fishes (of obvious importance to the Fish Commission) were all part of his scientific research.[15]

Throughout his long career at the Smithsonian Institution, Goode was able to use his scientific background to oversee acquisitions and scientific exchanges as the museum rapidly expanded its holdings. On his numerous trips to Europe, he inevitably visited major museums and often arranged for the transfer or loan of specimens for research or exhibition purposes, activities that required familiarity with materials and the needs of the museum.[16]

Goode, however, wrestled with questions regarding the significance of work in systematics and taxonomy near the end of the nineteenth century. Naturalists found themselves on the defensive regarding their methods of observation and analysis. They valued their tradition because field work made them familiar with context as well as detail. Darwin, they liked to point out, had been an observing naturalist whose theory of evolution was based on both intimate familiarity with specific species and a broad understanding of geographical context that required attention to detail beyond separate species. Periodic observation revealed changes over time, such as the elimination of some species and the intrusion of others. Older acquisitions held importance as reference material and were particularly valued when they were the original type specimens.

Louis Agassiz wrote, "In dealing with the history of the subject [of zoology], the value of each successive contribution should be estimated in the light of the knowledge of the period, not of that of the present time." [17] History, then, was an essential part of a naturalist's research and mode of thinking. Moreover, in the nineteenth century the public discovered history on a grand scale in ways that encompassed both natural history and human history.

III

Investigation, discovery, establishing connections, and documentation were not far removed from Goode's first major historical effort, a family genealogy. Knowledge of inheritance and environmental conditions was fundamental for those who worked in taxonomy and in genealogy, a comparability that was evident to many nineteenth-century naturalists. Still, somewhat embarrassed by this effort on family history, Goode confessed in the introduction to his massive volume entitled *Virginia Cousins: A Study of the Ancestry and Posterity of John Goode* that his research had been something of a family secret because genealogy was a study "not generally supposed to be of the highest order." [18]

Goode's historical research reinforced his conviction that, as in scientific theories of uniformitarianism and evolution, continuities were as evident in human society as in nature. Thus human history, from Goode's point of view, typically reflected slow and gradual change. Nonetheless, the amateur historian did not deny the exceptional importance of certain revolutionary, even catastrophic events—one of which was the impact of European arrival in the Americas on the native populations.

History, including natural history, was rediscovered and reworked in the nineteenth century. Grand visions and theories were paired with often painstaking taxonomical and archival work in an effort to establish a new sense of prehistory and human history. For Goode, the emphasis had been on the relatively rapid growth and theoretical changes within science, but as the end of the century neared he more and more resisted change and sought ways to direct it.[19] Goode was not, however, explicitly a philosopher or a social scientist. He argued that his approach was simply an examination of the historical record, like that of a field naturalist who described the terrain and the objects that seemed most evident and important. The past was to be investigated on its own terms, and yet described so as to demonstrate the order and significance determined by contemporary experts. He also emphasized the political and institutional environment in which individual activity was conducted.

Goode became increasingly interested in the history of science in the United States. His work on genealogy provided research tools that were useful as he gathered data for bibliographies and biographies of naturalists and physical scientists. Colonial history drew his initial attention, perhaps in part because Anglo-American connections had dominated that period.[20] Eventually, he focused his research on national scientific and educational institutions.[21]

Acquiring and organizing documentary evidence coincided with the contemporary movement in historical studies led by Herbert Baxter Adams from nearby Johns Hopkins University in Baltimore and given visibility through the American Historical Association, established in 1884. Under the guidance of a growing number of academics, history was to be scientific on the perceived German model, emphasizing documentation and objectivity. Samuel Scudder, an independent naturalist and briefly editor of *Science* magazine, approved of scientific history as he understood it: "The historian was to accompany the naturalist in his method of study, taking the thing to be studied in his hand, and applying the microscope to it; but this was to be done no longer with the ultimate principle of deducing general laws of human progress, but simply of completing the record."[22] While twentieth-century historians in a post-Beardian era are likely to be skeptical of such an innocent statement about objectivity, it is important to remember that Goode's generation was reacting against history as a romantic tale or a morality play.

Like other ubiquitous Washingtonians, Goode followed local political and intellectual initiatives carefully. He corresponded with Adams about the new American Historical Association and, when its act of incorporation passed Congress in 1889, the AHA was encouraged to deposit its collections, manuscripts, books and pamphlets at the Smithsonian. Not only did that make the Smithsonian its virtual headquarters, but the AHA also had its annual reports published under Smithsonian auspices (with a special Congressional appropriation). For the next two decades, the AHA and the Smithsonian were intimately connected. Goode was soon appointed to the AHA Executive Council and his assistant A. Howard Clark served as assistant secretary for the association. Goode's essay on "The Origins of the National Scientific and Educational Institutions of the United States" was originally presented to and published by the AHA.[23] Nor was the AHA Goode's only active historical connection, for he held offices in the local Columbian Historical Society and in the Sons of the American Revolution, the overshadowed sibling to the Daughters of the American Revolution.

Goode did not deny that there were growing distinctions between amateur and professional interests, whether in science, history, or museum work. He acknowledged and sought greater professionalism in each of those areas. At the same time, Goode believed that the experts had a responsibility to write and disseminate knowledge in ways accessible to a general public. He urged AHA members to publish more popular essays in its annual report. He encouraged museum curators to help educate the general public and chided those he identified as reclusive. Being a museum administrator provided Goode the opportunity to implement his ideas.

I V

Goode's commitment to examine in detail the development, functions, and goals of museums took hold during his Middletown years, and he had followed changing museum practices through the widely circulated publications of the Smithsonian Institution. The exceptional growth of its collections, as well as the simultaneous expansion of urban and collegiate museums throughout the United States and in Europe, made him aware of what most curators began to identify as the multiple and sometimes conflicting responsibilities for research, preservation, and public display. When he joined the Smithsonian in the 1870s, the museum operation was taking over much of its resources, functionally and intellectually. Publications, the exchange operation, and research were established components in the museum program of Joseph Henry when Goode arrived, but the growing number of collections to be housed and displayed required considerable discussion and space. What, if anything, was the value of the public exhibitions? If specimens were teaching devices, what should be taught? Equally important, what were effective teaching methods?

In the first half of the nineteenth century, natural history specimens had been acquired by private individuals, by entrepreneurs, by learned academies, by reform communities, by geological surveys, and by specialized study societies. These collecting activities had existed side by side, sometimes competing and sometimes reinforcing each others' efforts. By the 1850s, however, the practical difficulties of maintaining research museums had led to the decline of popular museums like that of Charles Willson Peale in Philadelphia and the virtual stagnation of many private and group study collections.[24]

In the last third of the nineteenth century, a number of large museums consolidated these resources and expanded the search for new materials in the United States and around the world. It was essential to redefine the function of museums. High ambitions led museum directors to lobby trustees, curators, and the public with their visions of grand public institutions.[25] Goode joined the international discussion with his *Circular Number One* published by the new United States National Museum. In it he argued that public museums had three distinct if interlocking functions. First and foremost, a *Museum of Record* would provide a safe repository for objects identified as "permanent land-marks of the progress of the world." An allied but distinctive component was to be a *Museum of Research,* which provided reference material for contemporary investigation and theoretical analysis. Third, an *Educational Museum* was fundamental to the implied public responsibility for instructing an educated citizenry. Classification and presentation of specimens in a cogent manner were essential. The tripartite scheme became the core of Goode's museum theory and its elements were elaborated in subsequent publications.

In the 1870s, curators at the Smithsonian agreed that the public displays should not be for mere amusement, a characteristic they criticized and equated with older antiquarian and eclectic museums. Goode insisted that exhibits provide the name, descriptive qualities, and scientific or cultural significance of every object. According to Goode, "an efficient educational museum may be described as a collection of instructive labels, each illustrated by a well-selected specimen."[26] The didactic approach was made more pressing by the fact that universal public education had become a national goal, object study was a mechanism increasingly used within the schools on nearly every level, poorly educated visitors sought explanations for what was displayed, and upper class patrons sought to shape expert knowledge for public consumption. There was also a demand for the better presentation of objects and, in the 1880s, more realistic preparations of bones and skins ultimately resulted in the dramatic habitat groups that distinguished turn-of-the-century museums of natural history.

In the 1870s and 1880s, along with dinosaurs, mastodons, and elephants, ethnology was a highly popular feature at international expositions and large museums. Since the 1850s Joseph Henry, who valued objects only when they could contribute to new knowledge, had regularly acquired archeological and ethnological artifacts. As Curtis Hinsley has argued, Henry followed then-current European theories about stages of human development and was eager to use American artifacts to contribute to the debates and changing interpretations.[27] The Smithsonian held extensive materials which grew steadily once the Bureau of Ethnology was housed in the Smithsonian. When Otis Mason and Goode were assigned to work on the Philadelphia Exposition, they used the opportunity not only to "show" Smithsonian resources but also to "tell" something of what experts understood about American natural history and ethnology.

The technique for teaching with artifacts was through arrangement and labels. Ideas were fundamental, and labels were the way to express and reinforce general themes in a larger exhibit. Goode came to critique public museums that lacked a meaningful arrangement and labels as simply "storehouses filled with the materials of which Museums are made." What was important were the concepts transmitted by specimens on display, ideas that related to current knowledge about science, technology, and human experience. Thus for Goode, "a finished museum is a dead museum." His *Classification of the Collection to Illustrate the Animal Resources of the United States* at the Exposition emphasized an organization of material in which the human experience was central and stressed the specific ways animals and other natural products could be utilized.[28]

In organizing materials on Native Americans at the National Museum,

Goode and his collaborator Mason found it difficult to choose between the two competing theories—one describing and organizing ethnological artifacts on a comparative and the other on an evolutionary basis. They eventually classified individual items in such a way that their system could encompass both theories. They recorded the physical character of artifacts (costumes, weapons, ceremonial items of particular groups) which allowed for a comparative approach. They also classified materials according to a particular stage (stone age, iron age, and so forth) in order to trace the evolution of culture and civilization without regard to race. Rather than commit the museum to one or the other approach, the pragmatic Goode had exhibit cases equipped with casters so that the entire display could be arranged either by function or developmental association. Objects of "mere curiosity" such as abnormal zoological specimens or miscellaneous historical items were relegated to storage.

Ethnology to some extent gave him an approach to the entire museum, using human agency as a fundamental organizing principle, while not obscuring the fact that much of nature remained beyond human influence. The National Museum displayed nature and also demonstrated what people had made of nature in the past, in other cultures, and what modern ingenuity had developed to date—that is, arts and industries of modern western civilization. The goal proved too comprehensive to be realized in Goode's administration, in part because he continued to rely heavily on volunteer staff and to pursue diverse themes within the museum. Nonetheless, it was Goode's genius that established the "purely educational element" of the Smithsonian.[29]

Goode's unusual capacity to attend to details and, at the same time, to address comprehensive issues made him increasingly a leader among museum administrators. In 1895, he was invited to address the British Museums Association at its annual meeting in Newcastle. He took the occasion to challenge his colleagues in museums and to codify "The Principles of Museum Administration." His keynote speech addressed issues raised in 1881 by William Stanley Jervons in an article on "The Use and Abuse of Museums" and stressed the importance of museums for popular education. Goode distinguished museums from universities and learned societies on the one hand and from expositions and fairs on the other, emphasizing that the museum in the late nineteenth century needed to provide "preservation of those objects which best illustrate the phenomena of nature and the works of man." [30] All materials should contribute to the increase of knowledge and provide for the culture and enlightenment of the people. Museum methods were of use in popular expositions and extended into zoological parks, botanical gardens, and public monuments; but their distinctive commitment to educating "the masses" made museums distinct. He argued to an already converted audience of professional administrators that leadership to maintain this responsibility needed to be carefully cultivated among museum staff. The curator was to use museum objects in order to make series of ideas intelligible, much as a printer could use the type in a printing office, letters and symbols capable of being used in literally hundreds of relationships. [31] The creative capacity of administrators to use the objects to illustrate forces and phenomena of nature would determine the educational influence of a museum. Goode wanted displays to illustrate contemporary ideas about the origin of specimens, as well as their development growth, function, structure and geographic distribution. Equally important, museums were obligated to reveal the relationship of objects to each other and their influence upon the structure of the earth and its human inhabitants. [32]

V

Goode's pioneering historical writings on science in America are synthetic, drawing together a wide array of literature and opinions in an effort to document the early and continuing attention to natural sciences.

Three long historical essays offer a comprehensive and unprecedented survey of people, events, and institutions from the beginning of settlement to Goode's own day. Two were presented before the Biological Society of Washington and one before the American Historical Association between 1886 and 1890. Trace elements of the naturalist, the historian, and the Washington administrator combine in a pioneering attempt to place science in the American setting of political democracy and diverse geography. Even as Goode wrote the essays, his emphasis also shifted, the naturalist losing ground to the administrator.

The first essay is entitled "The Beginnings of Natural History in America," and is typical of presidential addresses to scientific societies that acknowledge major contributors to an established tradition. It is innovative, however, in that Goode returns to the colonial period, recovering a past which, he argues, had "many men equal in capacity, in culture, in enthusiasm, to the naturalists of to-day, who were giving careful attention to the study of precisely the same phenomena of nature." Moreover, he discusses the difficult political and institutional environment in which they worked, relying on a tradition of voluntarism far from the European centers of knowledge. There was nothing predictable about the outcome of their efforts, according to Goode, but their intelligence and commitment made a contribution. Progress was possible but not inevitable. Thus, Goode argues that wars interrupted the work of science and, conversely, that significant promoters of science like Thomas Jefferson could advance it.[33]

His essay on colonial science and its sequel, "The Beginnings of American Science," brought together a cast of characters familiar to the biologists and elaborated on those who documented and classified American flora and fauna. These "intellectual ancestors" also laid foundations which undergird modern scientific structures. Goode reminds his audience that "without the encyclopædists and explorers there could have been no Ray, no Klein, no Linnæus. Without the systematics of the latter part of the eighteenth century, the school of comparative anatomy would never have arisen." Moreover, he notes, "certain institutions—museums, gardens, societies and government surveys—the germs . . . were in existence before the Revolution." Institutional development provides the framework and also the means to pursue science. Goode's history is surprisingly innocent: nowhere is there any concern about the nature of sources, the omission perhaps of certain participants, or the quality of documentation. What he does include is his own understanding of how politics can influence scientific work, concluding, for example, that radicalism could be disruptive. He deplores, for example, the sacrifices of time and the loss of productivity among the New Harmony naturalists of the 1820s who were pulled away from the intellectual and economic resources of Philadelphia to William Maclure's "socialist" utopia on the Wabash River in Indiana.

In his final historical survey, "Origin of the National Scientific and Educational Institutions of the United States," Goode not only discusses specific institutions but also probes their political dimensions. It was prominent scientists with political savvy, such as Alexander Dallas Bache, Joseph Henry, and Simon Newcomb, who developed plans for a scientific establishment in Washington with a research institute, an observatory, naturalists on exploring expeditions, and so forth. They built a strong cadre of institutions using the ideology and resources available in their own times unlike, he suggests, Joel Barlow, whose prospectus for a National Institution was unsuccessful because it resembled more closely the "House of Salomon in the New Atlantis of Bacon" than alternatives practical for the young United States.

Although sometimes seeming encyclopedic in its coverage, his narrative history carries an evident point of view very much in line with the relativism of contemporary historicists. Goode warms toward a theme of centralization and government sponsorship that reaches full significance as he completes his historical survey and arrives in the late nineteenth century. Using the Morrill land grant act as an example, he points out that government sponsorship had been more effective than informal advocacy in establishing agricultural colleges across the country. Independent scientific societies languished in many states, while state educational institutions—especially the colleges initiated by federal land grants—were becoming powerful forces for the advancement of science. Goode's history had reached a conclusion that coincided with his personal commitment to a policy of centralized sponsorship of educational and cultural institutions. His ideas were resonant with those progressives emerging with an agenda for social activism. There were perhaps similar motives in Goode's efforts to compile documents and write chapters for his massive history, *The Smithsonian Institution, 1846–1899.* The Smithsonian was a useful example, too, because both education and research benefited substantially from the consistent government sponsorship provided under Joseph Henry and Spencer Baird.[34]

Goode's history, then, had elements of public policy and an ideology that moved within the context of the contemporary historicism advocated by Adams and others but was not bound by it. Whether he was aware of the lively philosophical debates among historians, especially in Europe, is unclear. The essays show little explicit evidence of such influence, and reveal instead his naturalist's tendency to classify and to ascertain apparent relationships and patterns in the past that could account for the growth of science in the United States. His museum experience had helped him identify and address the issues that found an expression in his historical essays.

He began one of his essays with a quotation from Herbert Spencer: "Is not science a growth? Has not science, too, its embryology? And must not the neglect of its embryology lead to a misunderstanding of the principles of its evolution and of its existing organization?" George Brown Goode's historical writings were, in the final analysis, an effort to write a history so comprehensive and accurate that he forestalled such misunderstanding and helped establish an environment for future scientific productivity.

What he wrote was intimately connected to what he actually accomplished. Henry Fairfield Osborne remembered Goode as replete with "original ideas and suggestions, full of invention and of new expedients, studying the best models at home and abroad, but never bound by any traditions of system or of classification."[35] Goode's essay on "Museum-History and Museums of History" is an effort to explain how museums have been a fundamental part of past civilizations. He explains to fellow historians that there was still no comprehensive history museum on the model of those in natural history, but he argues that historical objects could also be a "most potent instrumentality for the promotion of historical studies." Once again, however, he warns against "a chance assemblage of curiosities" and emphasizes the importance of well-designed displays with clear labels.[36]

In his "Museums of the Future" Goode indulges his tendency to editorialize in a somewhat discursive account of museums, past and present. He welcomes the movement of museum development toward large, comprehensive public institutions, accessible and attractive to the public. The model museum for the future should be a "house full of ideas, arranged with strictest attention to system."

Surrounded by colleagues in Washington whose efforts forecast a progressive outlook, Goode stressed the opportunities museums could provide. Museums as repositories of specimens and information were a place for experts to gather and analyze the data in relationship to contemporary intellectual and practical concerns. Moreover, they were essential for public education. He envisioned museums of the future as "chief agencies of higher civilization" and likened them to public libraries and other urban cultural institutions under development at the end of the nineteenth century.

His commentary struck a responsive chord among a new generation of aspiring museum professionals. He represented the new professional administrator and, like others of his generation, delighted in telling tales about past museum practice. For Smithsonian personnel, the fate of natural history materials that had been housed in the Patent Office was familiar. With wry humor Goode recounted one story of a Commissioner of Patents who, annoyed by the collection of fossil vertebrates filling his crowded space, sent them to a bone mill in Georgetown. There, he pointed out, they were converted into commercial fertilizers—"once food for thought, they now became food for the farmers' crops."[37] The new professionals sought to ground their museum practice in stated policies to prevent such losses, and to make museum administration a place for discussion of museum purpose in theory as well as in practice.

Goode's visions were tempered with a kind of realism about what was possible to accomplish in his generation. The transitional nature of his efforts restrained him from undue pride and required an open-ended outlook in which process was as important as content. The historical essays reprinted here are a reconnaissance over historical territory which Goode anticipated would be explored in greater detail by later scholars. Nonetheless, he drew his own conclusions, reassessing, for example, the reputations of earlier naturalists and concluding that the eccentric Constantine Rafinesque had been defamed and suggesting that William Bartram had been as much underrated as John Bartram had been unduly exalted. He also stressed the importance of minor collectors and encyclopedists whose work was fundamental to later theorists. The gradual decline in the natural history tradition and the growing emphasis on specialization and experimentation in his own generation was a cause for concern, primarily because it left investigators "strangely indifferent to the question as to how the public at large is to be made familiar with the results of their labors."[38] Goode's personal career trajectory had, of course, been away from specialization and toward the dissemination of knowledge. As a museum curator and director he had, with unusual vigor and insight, worked to establish a theoretical base from which to address the practical problems of presenting natural history in a public museum.

Perhaps it is appropriate to conclude with a personal note. My interest in Goode goes back a long time, to the years when I was finishing a dissertation and came across Goode's historical essay on the National Institute for the Promotion of Knowledge. I then read the Institute's own records at the Smithsonian Institution, many of them acquired long after Goode's administration. With the brashness of a doctoral student, I took on his interpretation of the Institute as a precursor of the American Association for the Advancement of Science, on which I was then working, and of the Smithsonian. This research resulted in an article in *Isis* that challenged Goode's interpretation and pointed out the fundamental weaknesses of the Institute as a scientific organization.[39] Surely the fates are amused by this editing project. My research on Goode and on museum history has me rereading records of the Institute, interpreting in new ways the hostile outlook of contemporary scientists, and understanding why Goode was so insistent on crediting its members' prescient efforts to permit public access to science. I have acquired a new respect for Goode as a pioneering historian and have a clearer understanding about the differences in our interpretations of such nineteenth-century phenomena.[40]

THE BEGINNINGS OF NATURAL
HISTORY IN AMERICA[1]

By George Brown Goode

President of the Biological Society of Washington

Is not science a growth? Has not science, too, its embryology? And must not the neglect of its embryology lead to a misunderstanding of the principles of its evolution and of its existing organization?

—Spencer: *The Genesis of Science.*

ANALYSIS.

I. Thomas Harriot, the earliest English naturalist in America
II. Harriot's Spanish and French predecessors and contemporaries
III. Garcilasso de la Vega and the biological lore of the native Americans
IV. Anglo-American naturalists of the seventeenth century
V. European explorations in the New World, 1600–1800
VI. The founders of American natural history
VII. The debt which the naturalists of the present owe to those of the past

I

Three centuries ago the only English settlement in America was the little colony of one hundred and eight men which Raleigh had planted five months before upon Roanoke Island, in North Carolina.

The 17th of August, 1885, was the anniversary of one of the most noteworthy events in the history of America, for it marked the three hundredth return of the date when Sir Richard Grenville brought to its shores this sturdy company of pioneers, who, by their sojourn on this side of the Atlantic, prepared the way for the great armies of immigrants who were to follow.

It was also the anniversary of an important event in the history of science, for among the colonists was Thomas Harriot, the first English man of science who crossed the Atlantic. His name is familiar to few save those who love the time-browned pages and quaint narrations of Hakluyt, Purchas, and Pinkerton; yet Harriot was foremost among the scholars of his time—the Huxley or the Stokes of his day—a man of wide culture, a skillful astronomer, a profound mathematician, the author of a standard treatise upon algebra, and a botanist, zoologist, and anthropologist withal. "He had been the mathematical instructor of Raleigh, and in obeying this summons to go forth upon the present expedition gave to it," says Anderson, "the most valuable aid which could be derived from human strength." [2]

This eminent man deserves more than a passing notice on this occasion, and I have taken pains to bring together all that is known about him. He was born at Oxford in 1560, or, as old Anthony Wood quaintly expresses it, "he tumbled out of his mother's womb into the lap of the Oxonian muses," and at an early age was entered as a scholar in St. Mary's Hall, receiving his bachelor's degree in 1579. He was soon received into Raleigh's family as his instructor in mathematics, and at the age of twenty-five made his voyage to America.

After his return he was introduced by Raleigh to Henry Percy, Earl of Northumberland, one of the most munificent patrons of science of that day, who allowed him a pension of £120 a year. "About the same time," we are told, "Hues, well known by his Treatise upon the Globes,[3] and Walter Warner, who is said to have given Harvey the first hint concerning the circulation of the blood, being both of them mathematicians, received from him (Northumberland) pensions of less value; so that in 1606, when the Earl was committed to the Tower for life, Harriott, Hues, and Warner were his constant companions, and were usually called the Earl of Northumberland's Magi." [4]

One thing, at least, have three centuries accomplished for science. Its greatest workers are not now, as they were at the beginning of the seventeenth century, dependent upon the liberality and caprice of wealthy men, classed as their "pensioners" and "servants," and assigned places at their tables which they must needs accept or famish.

Harriot appears to have passed the latter years of his life at Sion College, near Isleworth, where he died in 1621. He was buried in St. Christopher's Church, London, and the following eulogy was embodied in his epitaph:

QUI OMNES SCIENTIAS CALLUIT AC IN OMNIBUS EXCELLUIT
MATHEMATICIS, PHILOSOPHICIS, THEOLOGICIS,
VERITATIS, INDAGATOR STUDIOSISSIMUS,
DEI TRINIUNIUS PIISSIMUS.

He was especially eminent in the field of mathematics. "Harriott," says Hallam, "was destined to make the last great discovery in the pure science of algebra. . . . Harriott arrived at a complete theory of the genesis of equations, which Cardan and Vieta had but partially conceived."[5]

His improvements in algebra were adopted, we are told, by Descartes, and for a considerable time imposed upon the French as his own invention, but the theft was at last detected and exposed by Doctor Wallis in his Treatise of Algebra, both Theoretical and Practical, London, 1685.[6]

"Oldys, in his Life of Sir Walter Raleigh, has shown," says Stith, "that the famous French philosopher, Descartes, borrowed much of his light from this excellent mathematician, and that the learned Doctor Wallis gave the preference to Hariot's improvements before Descarte's, although he had the advantage of coming after and being assisted by him."[7]

Harriot's papers were left after his death in the possession of the Percy family at Petworth, where they were examined in 1787 by Doctor Zach, and later by Professor Rigaud, of Oxford, who, in 1833, published in his supplement to the works of James Bradley, An Account of Thomas Harriot's Astronomical Papers. His observations on Halley's comet in 1607 are still referred to as being of great importance. Zach pronounced him an eminent astronomer, both theoretical and practical. "He was the first observer of the solar spots, on which he made a hundred and ninety-nine observations; he also made many excellent observations on the satellites of Jupiter, and, indeed, it is probable that he discovered them as early if not earlier than Galileo."[8]

A posthumous work, Artes Analyticæ Praxis ad Æquationes algebraicas nova, expedita et generali Methodo resolvendas, e posthumis Thomas Harriot, was published in 1631 by his friend and associate, Walter Warner, and there is in the library of Sion College a manuscript work of his entitled Ephemeris Chyrometrica.

Wood says that, "notwithstanding his great skill in mathematics, he had strange thoughts of the Scriptures, always undervalued the old story of the creation of the world, and would never believe that trite proposition, 'Ex nihilo nihil fit.'"

Stith, the historian of Virginia, protests, however, against the charge that Harriot had led his pupil Raleigh into atheism. "As to this groundless Aspersion," he remarked, "the Truth of it, perhaps, was that Sir *Walter* and Mr. *Hariot* were the first who ventured to depart from the beaten Tract of the Schools, and to throw off and combat some hoary Follies and traditionary Errors which had been riveted by Age, and rendered sacred and inviolable in the Eyes of weak and prejudiced Persons. Sir *Walter* is said to have been first led to this by the manifest Detection, from his own Experience, of their erroneous Opinions concerning the *Torrid Zone;* and he intended to have proceeded farther in the Search after more solid and important Truths 'till he was chid and restrained by the Queen, into whom some Persons had infused a Notion that such Doctrine was against God."[9]

The erroneous opinions concerning the torrid zone which were called in question by Harriot and Raleigh were based upon a statement of Aristotle, in those days accepted as an article of faith, that the equatorial zone of the earth was so scorched and dried by the sun's heat as to be uninhabitable. Even the experience of explorers was for many years overpowered by the weight of this time-worn dogma. The Jesuit, Acosta, was accused of atheism on the same grounds by his Spanish contemporaries, but he rejoiced that he had seen for himself and that the climate under the equator was so different from what he had expected that "he could but laugh at Aristotle's meteors and his philosophy."

Harriot's Brief and True Report of the New Found Land of Virginia, a thin volume in quarto, printed at Frankfort on the Main in 1590,[10] is now one of the rarest and most precious works relating to America[11] and is full of interest to the naturalist. Harriot's description of the Indians and their customs and beliefs, though strongly tinctured with prepossessed ideas concerning them, is thorough and scholarly, and one of the fullest and most reliable of the early treatises upon the inhabitants of North America.

The chief man of the Roanoke colony, Sir Ralph Lane, usually spoken of as the first governor of Virginia, was a man of great energy and enterprise,[12] and with the help of Harriot planned and conducted expeditions in every direction—southward, 80 leagues to Secotan, "an Indian town, lying between the rivers Pampticoe and Neus;" to the northwest, up the Albemarle Sound and Chowan River to the forks of the Meherrin and Nottaway; and north, 130 miles to the Elizabeth River, on the south side of Chesapeake Bay.

Besides his description of the Indians, Harriot wrote "a particular narrative of all the beasts, birds, fishes, fowls, fruits, and roots, and how they may be useful." A systematic report could hardly be expected from one who lived a century and a half before Linnæus, but if we keep in mind the condition of zoology at that day we can but be pleased with the fullness of his narrative.

He collected the names of twenty-eight species of mammals, twelve of these, including the black bear, the gray squirrel, the cony or hare, the otto, and the possum and raccoon (*Saquenúckot* and *Maquówoc*), he saw, beside the civet cat or skunk, which he observed by means of another sense. He was the first to distinguish the American from the European deer, stating that the former have longer tails, and the snags of their horns look backward—a brief diagnosis, but one which was not replaced by a better one for nearly two centuries.

Of birds he collected the names of eighty-six "in the countrie language," and had pictures drawn of twenty-five. He mentions turkeys, stockdoves, partridges, crows, herons, and, in winter, great store of swans and geese.

With aquatic animals he seems to have been well acquainted. He refers to some by English names, and to many others which had no names "but in the countrey language." In the plates accompanying the first edition of his book are figured several familiar forms, then for the first time made known in Europe, among them the gar pike (*Lepidosteus*),[13] and the horse shoe or king crab (*Limulus*),[14] "*Seekanauk*, a kinde of crustie shell fishe which is good meate, about a foot in breadth, having a crustie tayle, many legges like a crabbe, and her eyes in her back."

Harriot also alludes to various kinds of trees and shrubs, usually by their Indian names. Among them may easily be recognized the pitch pine, sassafras, shoemake, chestnut, walnut, hickory, persimmon, prickly pear, Nelumbium, Liriodendron, holly, beech, ash, and so on, beside the maize and tobacco cultivated by the natives.

A companion of Harriot's, whose labors are deserving of notice, was John With or White, the first delineator of plants and animals who visited this continent. Concerning him and the ultimate utilization of his work, Stith discourses as follows:

UPON this Voyage, Sir *Walter Raleigh*, by the Queen's Advice and Directions, sent, at no small Expence, Mr. *John With*, a skilful and ingenious Painter, to take the Situation of the Country, and to paint, from the Life, the Figures and Habits of the Natives, their Way of Living, and their several Fashions, Modes, and Superstitions; which he did with great Beauty and Exactness. There was one *Theodore de Bry*, who afterwards published, in the Year 1624, the beautiful *Latin* Edition of Voyages, in six Volumes, *Folio*, a most curious and valuable Work. He being in *England* soon after, by the Means of the Rev. Mr. *Richard*

Hackluyt, then of *Christ's-Church*, in *Oxford*, who, *De Bry* tells us, had himself seen the Country, obtained from Mr. *With* a Sight of these Pieces, with Permission to take them off in Copper Plates. These, being very lively and well done, he carried to *Frankfort*, on the *Maine*, where he published a noble Edition of them, with *Latin* Explanations, out of *John Wechelius's* Press, in the Year 1590. And these are the Originals from which Mr. *Beverley's*, and the Cuts of many of our late Writers and Travellers, have been chiefly imitated.[15]

With's drawings are still in the British Museum,[16] where they were examined in 1860 by Doctor E. E. Hale, who reported upon their condition to the American Antiquarian Society.[17]

This collection, he says, consists of one hundred and twelve drawings in water color, very carefully preserved. They are very well drawn, colored with skill, and even in the present state of art would be considered anywhere valuable and creditable representations of the plants, birds, beasts, and men of a new country. Mr. Hale gives a list of these drawings as identified by Sloane and others. Among these were the bald eagle, the red-headed, hairy, and golden-winged woodpeckers, the blue-bird, red-wing blackbird, towhee, redbird, blue jay, and fox-colored thrush, the crow blackbird, and apparently the mocking bird—"*Artamockes*, the linguist; a bird that imitateth and useth the sounds and tones of almost all birds in the countrie." Among the fish we recognize the mullet (*Tetzo*), the menhaden or oldwife (*Masunnehockeo*), and the sturgeon (*Coppauleo*), and perhaps the squeteague or chigwit (*Chigwusso*).

The science of North America, then, began with Thomas Harriot. Let us review together to-night its progress for a period of two centuries—a period coinciding almost exactly with the colonial portion of the history of the United States.

"The present generation," says Whewell, "finds itself the heir of a vast patrimony of science, and it must needs concern us to know the steps by which these possessions were acquired and the documents by which they are secured to us and our heirs forever. Our species from the time of its creation has been traveling onward in pursuit of truth; and now that we have reached a lofty and commanding position, with the broad light of day around us, it must be grateful to look back on the line of our past progress; to review the journey begun in early twilight amid primeval wilds, for a long time continued with slow advance and obscure prospects, and gradually and in later days followed along more open and lightsome paths, in a wide and fertile region. The historian of science, from early periods to the present time, may hope for favor on the score of the mere subject of his narrative, and in virtue of the curiosity which the men of the present day may naturally feel respecting the events and persons of his story."

II

Although Harriot was the first who described the natural characteristics of North America, it would not be proper to ignore the fact that the first scientific exploration of the Western Continent was accomplished by Spaniards and Frenchmen.

Gonzalo Fernandez de Oviedo y Valdes, the first historian of the New World [b. 1478, d. 1557], was an Asturian of noble birth, who began life as a page in the palace of Ferdinand and Isabella. He saw Columbus at Burgos on his second return from America in 1496. He came over in 1514 to Santo Domingo, having been appointed inspector of gold smelting, and was subsequently governor of that island and royal historiographer of the Indies. In 1525 he transmitted to Charles V his Sumario de la Natural Historia de las Indias, printed at Toledo two years later, and in 1535 began the publication of his Historia Natural y General de las Indias, a task which was finally completed only thirty years ago by the Spanish Royal Academy of History.

Las Casas said that Oviedo's books were "as full of lies almost as pages," but whatever may have been his methods in the discussion of history and politics, he seems, in his descriptions, to have been both minute and accurate. Among the American animals which he was first to mention was the tapir or *dant*—"of the bignesse of a meane mule, without hornes, ash-coloured," and the *churchia*, evidently a species of *Didelphys*, allied to our possum. This was the first notice of any member of the great group of marsupial mammals. I quote a portion of the description in Oviedo's Sumario, employing the quaint phraseology of Purchas's translation:

The *Churchia* is as bigge as a small *Conie*, tawnie, sharpe-snowted, dog-toothed, long-tayled and eared like a *Rat*. They do great harm to their Hennes, killing sometimes twenty or more at once to sucke their bloud: And if they then have young, shee carrieth them with her in a bagge of skin under her belly, running alongst the same like a Satchell, which shee opens and shuts at pleasure to let them in and out.[18]

He characterized and described at length many other animals, among them the manatee, the iguana (*Iuanna*), the armadillos (*Bardati*), the anteaters, the sloth, the pelican, the ivory-billed woodpecker, and the humming birds.

There are found in the firme land [he wrote] certaine birds, so little that the whole bodie of one of them is no bigger then the top of the biggest finger of a mans hand, and yet is the bare body without the feathers not half so bigge. This Bird, besides her littlenesse, is of such velocitie and swiftness in flying, that who so seeth her flying in the aire, cannot see her flap or beate her wings after any other sort then doe Dorres, or the Humble Bees, or Beetles. . . . . And I know not whereunto I may better liken them, then to the little birds which the lymners of bookes are accustomed to paint on the margent of Church Bookes, and other Bookes of Divine Service. Their Feathers are of manie faire colours, golden, yellow, and greene.

That the spirit of Oviedo's work was scientific and critical, and not credulous and marvel-seeking, like that of many of his contemporaries, is everywhere manifest. His materials are classified in systematically arranged chapters. His methods may be illustrated by referring to his chapter On tigers.

"In Terra Firma," he begins, "are found many terrible beasts which the first Spaniards called tigers—which thing, nevertheless, I dare not affirm." He then reviews concisely and critically what is known of tigers elsewhere, and goes on to describe the supposed American tiger at length, and in such terms that it is at once evident that the mammal under discussion is one of the spotted cats, doubtless the jaguar (*Felis onca*).[19]

The second in order of time to publish a book upon American natural history was Jean de Lery [b. 1534, d. 1611], a Calvinistic minister, who was a member of the Huguenot colony founded by the Chevalier de Villegagnon in 1555, on the small island in the bay of Rio de Janeiro, which still bears his name. He remained in Brazil less than five years, and in 1578 published at Rouen a work entitled Voyage en Amerique, avec la description des Animaux et Plantes de ce Pays.

Joseph d'Acosta was another Spanish explorer who preceded Harriot, and was a man of much the same school and temper of mind. Born in the province of Leon about the year 1539, he entered the society of Jesuits at the age of fourteen, and in 1571 went to Peru, where he traveled as a missionary for seventeen years. After his return to Spain he filled several important ecclesiastical offices and died February 15, 1600, rector of the University of Salamanca. His first book, De Natvra Novi Orbis Libri dvo, was published in 1589. His Historia Natvral y Moral delas Indias appeared in 1590, and is one of the best known and most useful of the early Spanish works on America, having passed through numerous editions in many languages.

Acosta was, perhaps, the most learned of the early writers upon America, and his writings, though modeled after those of the mediæval schoolmen, were full of suggestive observations, "touching the naturall historie of the heavens, ayre, water, and earth at the West Indies, also of their beasts, fishes, fowles, plants, and other remarkable varieties of nature." He discoursed "of the fashion and form of heaven at the new-found world," "of the ayre and the winds," of ocean physics, of volcanoes and earthquakes, as well as of metals, pearls, emeralds, trees, beasts, and fowls.

He discussed the appearance and habits of the manatee and the crocodile, and described the Indian methods of whaling and pearl fishing. He dwelt at length upon the condition of the domestic animals, sheep, kine, goats, horses, asses, dogs, and cats which the Spaniards had introduced into the New World and which were already thoroughly acclimated. It seems strange to learn from his pages that in the year 1587, 99,794 hides of domestic cattle were exported from Santo Domingo and New Spain to Seville. Lynceus has suggested that some of these skins were from the bison herds, believed at that time to have been abundant in the north of Mexico.

He gives a formidable catalogue of the animals of Central and South America, in which occur the familiar names of armadillo, iguana, chinchilla, viscacha, vicugna, paco, and guanaco, and describes many of them at length, especially the peccary (*Saino*), the tapirs, the sloths, and the vicugna. He speaks of the cochineal insect, which had already become of importance in the arts.

He was the first to call attention to the existence in South America of immense fossil bones; these he supposed to be the remains of gigantic individuals of the human species.

His description of the flora is very full, and he dwells at length upon the useful applications of the cacao bean and its product, the drink which they call chocolate—"whereof they make great account in that country, foolishly and without reason"—the plantain, the yucca, the cassava, the maguey, the tunall or cactus, and very many more.

It is, however, as a scientific theorist that Acosta has the highest claim to our attention. He appears to have been the first to discuss America from the standpoint of the zoogeographer.

In considering the question, "How it should be possible that at the Indies there should be any sorts of beasts, whereof the like are nowhere else," he owns that he is quite unable to determine whether they were special creations or whether they came out of the ark. He evidently prefers the first alternative, although so trammeled by the prevalent opinions of his day and sect that he is unable to bring himself quite to its avowal. He approaches so close to the limits of heterodoxy, however, that Purchas, in His Pilgrimmes, feels obliged to print a footnote, pronouncing it "un-Christian to say that America was not drowned with the flood."

Acosta thoroughly appreciated the peculiar character of the American fauna, and remarked that "if the kinds of beasts are to be judged by their properties, it would be as reasonable to call an egg a chestnut as to seeke to reduce to the known kinds of Europe the divers kinds of the Indies." He was even willing to admit that it may not be necessary to say that the creation of the world was finished in six days, and that beasts of a more perfect character may have been made subsequently; and in his anxiety to escape the alternative of a Noah's ark almost committed himself to a theory of evolution. "We may consider well upon this subject," he wrote, "whether these beasts differ in kinde and essentially from all others, or if this difference be accidentall, which might grow by divers accidents, as we see in the Images of men, some are white, others black, some Giants, others Dwarfes; and in Apes, some have no taile, others have; and in Sheepe, some are bare, others have fleeces, some great and strong with a long necke, as those of Peru, others weake and little, having a short necke, as those of Castile. But to speak directly, who so would preserve the propagation of beasts at the Indies and reduce them to those of Europe, hee shall undertake a charge he will hardly discharge with his honour."

Francesco Hernandez, a representative physician and man of science, was sent by Philip II of Spain to Mexico, with unlimited facilities for exploration, and remained in that country from 1593 to 1600. His notes and collections seem to have been very extensive, and it is said that over 1,200 drawings of plants and animals were prepared under his direction. Editions of his works were published in Mexico in 1604 and 1615. I am assured by Mexican naturalists that his work was careful and valuable, the only defect being that he trusted too implicitly in what he was told by the native Mexicans.

Among the animals not met with in previous writings are the coyote (Aztec, *Coyotl*), the buffalo, the axolotl, the porcupine (*Hoitztlacuatzin*), the prong-buck (*Mazame*), the horned lizard (*Tapayaxin*), the bison, the peccary (*Quapizotl*), and the toucan.

Among those of which figures are for the first time published are the ocelot (*Ocelotl*), the rattlesnake (*Teuhtlacot zanhqui*), the manatee (*Manati*), the alligator (*Aquetzpalin*), the armadillo (*Ayotochtli*), the pelican (*Ayototl*).

The figures of plants are numerous, and in most instances, I should judge, recognizable.

Many other Spaniards published their observations upon America in the sixteenth and seventeenth centuries, but it is perhaps not necessary to refer to them even by name. They were, as a rule, travelers, not explorers. Purchas assures us that "Acosta and Oviedo have best deserved of the studious of Nature—that is, of the knowledge of God in his workes."

III

A personage who must on no account be overlooked in the consideration of these early days is Garcilasso de la Vega. Born in Peru in 1539, his father the Spanish governor of Cuzco, his mother a princess of the Inca blood, he boasted of a lineage traced through the line of ancient Peruvian monarchs back to Manco Capac and the Sun. He served as a soldier in

Europe and died in Spain about the year 1617. His Royal Commentaries of Peru, constitutes a magnificent contribution to the history of pre-Columbian America, and was said by some authorities to have been first written in the Peruvian language.[20]

Be this as it may, De la Vega's commentaries, though more valuable to the civil than to the natural historian, will always possess a peculiar interest, not only because the author was the first native of America who wrote concerning its animals and plants, but for the reason that it represents to us the historic and scientific lore of the aboriginal inhabitants of this continent.

De la Vega describes in an intelligible manner the condor (*Cuntur*) of South America, of which, as he tells us, there was a famous Indian painting in the temple at Cacha, the mountain cats or ocelots (Inca *Ozcollo*, Aztec *Ocelotl*), the puma, the viscacha, the tapir, and the three-toed ostrich. He was one of the first to notice the skunk (*Mephitis*, sp.), "which the Indians call Annas, the Spanish *Zorinnas*." "It is well," he remarks, "that these creatures are not in great numbers, for if they were, they were able to poison and stench up a whole countrey." He devotes a chapter to "the tame cattel which God hath given to the Indians of Peru"— the llama and the huanaco—and speaks also of the paco and the vicuna, clearly distinguishing and describing the appearance and habits of the four species of Tylopoda which occur on the west coast of South America, although European naturalists a century later knew but two of them. He describes the annual vicuna hunts which were conducted by the Inca kings in person, assisted by twenty or thirty thousand Indians.

The fauna of Peru, as catalogued by him, included nearly fifty species, and the minuteness of his observations and the accuracy of his descriptions are very surprising. He discusses at length the plants of Peru, especially the maguey, the pineapple, the tobacco, and the "pretious leaf called *Cuca*," whose virtues pharmacologists now hold in such high esteem, and devotes chapters to "The Emeralds, Turquoises, and Pearls of that Countrey;" to gold and silver, and to quicksilver.

De la Vega refers to a certain place in the city of Cuzco, where lions and other fierce creatures are kept in captivity. The taste for menageries and gardens seems to have been less pronounced in Peru, however, than in Mexico.

Much has been written concerning the wonderful collection of animals and plants which the Spanish conquistadors found in Montezuma's capital city. Carus, in his Geschichte der Zoologie declares that at the time of the discovery of Mexico, Europe had no menageries and botanical gardens which could be compared with those of Chapoltepec and Huextepec, a statement which is quite within the bounds of truth, for the earliest botanical garden in the Old World was that founded at Pisa in 1543.[21] Our fellow member, Doctor Charles Rau, has also described the zoological gardens of Mexico in glowing terms,[22] and Professor E. B. Tylor states that in the palace gardens of Mexico all kinds of birds and beasts were kept in well-appointed zoological gardens, where there were homes even for alligators and snakes, and declares that this testifies to a cultivation of natural history which was really beyond the European level of the time.

Is it not to be regretted that the capital of the United States in 1885 is still unprovided with a means of public instruction which was to be found in the capital of Mexico four hundred years ago?

I have examined the historians of Mexico with care, and must express my conviction that the truth is more nearly touched in the bluff, soldier-like narrative of Cortez himself than in the flowery and redundant paraphrases of Prescott. We may, probably, safely accept the story as told by Bernal Diaz del Castillo, one of the companions of Cortez, to whom Torquemada, Robertson, Lockhart, Rau, and others give high praise as a truthful narrator.

Diaz presents a most vivid word-painting of the city of Mexico, and was particularly impressed by the royal aviaries:

We saw here every kind of eagle, from the king's eagle[23] to the smallest kind included; and every species of bird, from the largest known to the little colibris,[24] in their full splendor of plumage. Here also were to be seen those birds from which the Mexicans take the green-colored feathers, of which they manufacture their beautiful feathered stuffs These last-mentioned birds very much resemble our Spanish jays and are called by the Indians *quezales*.[25]

The species of sparrows[26] were very curious, having five distinct colors in their plumage—green, red, white, yellow, blue.

There were such vast numbers of parrots and such a variety of kinds that I can not remember all their names; and geese of the richest plumage and other large birds.

These were at stated periods stripped of their feathers, that new ones might grow in their place. All these birds had appropriate places to breed in and were under the care of several Indians of both sexes, who had to keep their nests clean, give to each kind its proper food, and set the birds for breeding.

In another place, near a temple, were kept all manner of beautiful animals, the names of which were not noted by Diaz, nor their peculiarities described.

In the building where the human sacrifices were perpetrated there were dens in which were kept poisonous serpents, and among them "a species at the end of whose tail there was a kind of rattle." This last-mentioned serpent, which is the most dangerous, was kept in a cabin in which a quantity of feathers had been strewed; here it laid its eggs, and it was fed with the flesh of dogs and of human beings which had been sacrificed. . . . When all the tigers and lions[27] roared together with the howlings of the jackals[28] and foxes and hissing of the serpents, it was quite fearful, and you could not suppose otherwise than that you were in hell.

This is the first record of the rattlesnake, and brings to mind the captive snakes of the Mokis, their annual snake dance, and their use of feathers in the same connection.[29]

I am not yet prepared to believe in the marvelous aquaria described by Prescott, although fish ponds there doubtless were.

I am assured by our fellow-member Señor Aquilera, that the locations of the gardens of Montezuma are well identified, and that the Mexican Indians still possess a marvelous knowledge of the medicinal virtues of plants, which is handed down by tradition from generation to generation. From this he infers that in the days of Aztec glory the knowledge of the uses of plants must have been very comprehensive.

Who shall say that the spirit of true science did not inspire the Inca Pachacutec, when many centuries ago he handed down to his descendants maxims such as this:

A herbalist who knows the names but is ignorant of the virtues and qualities of herbs, or he who knows few but is ignorant of most, is a mere quack and mountebank, and deserves not the name and repute of a physician until he is skillful as well in the noxious as in the salutiferous qualities of herbs.

Impressed with the extent of the knowledge of nature among the aborigines of America, I asked one of the most learned of our anthropologists for his opinion in regard to its character, and received the following statement:

WASHINGTON, *January 5, 1886.*
MY DEAR MR. GOODE: We make a very grave mistake if we think there was no study of nature before the science of natural history. In all branches of study whatever there was lore before there was science. Before the Weather Bureau was weather lore, a kind of rough induction which the ancient people made, and which was very far from erroneous, Doctor Washington Matthews read a paper before the Washington Philosophical Society more than a year ago[30] to draw attention to the marvelous intimacy of the Navajo Indians with the plant kingdom around them, and their vocabulary, which contained names for many species constructed so as to connote qualities well known to them. You are familiar with the stories concerning the respect in which certain animals are held by the Eskimo, and the minute acquaintance of all our aborigines of both

continents with the life histories of many animals. The Eskimo, as well as the Indian tribes, carve and depict forms so well that the naturalist can frequently determine the species. Mr. Lucien Turner collected carvings in ivory of fœtal forms.

Very truly, yours,   O. T. MASON.

Professor Mason also called attention to a long paper upon Tame Animals among the Red Men of America, by Doctor E. F. im Thurn,[31] in which it is stated that the Indian of South America finds means to tame almost every wild bird and beast of his country, so that these domesticated animals are ever among the most prominent members of his household, not because of any affection for them, but because he enjoys their bright colors, makes use of them in various ways, and employs them as a medium of exchange. They even know how to change the colors of a living bird from green to yellow. In one settlement he counted twenty-one kinds of monkeys. Nearly all of the thirty or more species of Guiana parrots are tamed, two species of deer, two of peccaries, two of coatimundis, jaguars, pacas, capybaras, agoutis, hawks, owls, herons, plovers, toucans, troupials, rupicolas, and iguanas were also observed in captivity. The mere fact that these animals are kept in captivity is not in itself especially significant, but it renders it possible to understand how splendor-loving rulers of Mexico succeeded in building up the great menageries.

Bearing in mind the animal myths which Major Powell has found so prevalent among the Indians of Arizona and New Mexico, and has so charmingly translated, and those which Schoolcraft and others recorded in the north long ago, and which Longfellow has arranged in metric form, we can not but be impressed with the idea that the red man of old, living close to nature as he did, knew many of her secrets which we should be glad to share with him at the present day.

Garcilasso de la Vega was not the only descendant of the aboriginal Americans who has written upon their history. Among the authors of works upon Mexican archæology published in the seventeenth and eighteenth centuries were Taddeo de Niza and Gabriel d'Ayala, "noble Indians" of Tlazcala and Tezcuco, the three named Ixtlilxochitl, and ten or twelve more. Gongora, a native Mexican, professor of mathematics in the University of Mexico, was one of the earliest American astronomers, the author of the Mexican Cyclography, printed two centuries ago. Herrera, Martinez, Garcia, Torquemada, Castillejo, De Betancourt, De Solis, Del Pulgar, and Beneducci have done what they could to preserve a portion of this ancient American lore, and it seems almost incredible that, sometime in the future when American archæology shall have gained a firmer footing, some of the treasures of fact which these men garnered up are not to have an important function in elucidating anthropological problems which are as yet entirely unsolved.

IV

The colony on Roanoke Island having been abandoned by the English, twenty years elapsed before their next effort toward peopling America. Then came the adventurers to Jamestown in 1606, and with them that picturesque personage, Captain John Smith, who, though unversed in the mathematics and astronomy which made up to a great extent the science of the day, was a keen observer and an enterprising explorer. His contributions to geography were important, and his descriptions of the animals and plants of Virginia and New England supplement well those of his predecessor, Harriot.

Captain Smith was the first to describe the raccoon, the musquash, and the flying squirrel:

There is a beast they call *Aroughcun* (raccoon), much like a badger, but useth to live on trees, as Squirrels doe. Their Squirrels some are neare as great as our smallest sort of wilde Rabbets, some blackish, or blacke and white, but most are gray. A small beast they have they call *Assapanick*, but we call them flying Squirrels, because, spreading their legs, and so stretching the largenesse of their skins that they have been seene to fly 30 or 40 yards. An *Opossum* hath a head like a Swine, and a taile like a Rat, and is of the bignesse of a Cat. Vnder her belly she hath a bagge, wherein she lodgeth, carrieth, and suckleth her young. A *Mussascus* (musquash) is a beast of the forme and nature of our water Rats, but many of them smell exceedingly strongly of Muske.

And in the same strain he goes on to mention a score of mammals, identifying them with those of Europe with surprising accuracy.

His "*Utchun quoyes*, which is like a Wild Cat," is evidently the bay lynx. With the birds he was less familiar, but he mentions a number which resemble those of Europe, and states that many of them were unfamiliar. He was the first to refer to the red-wing blackbird (*Agelæus phœniceus*).

He catalogues twenty-five kinds of fish and shellfish, using the names by which many of them are known to this day.

He gives also a very judicious account of the useful trees of Virginia, referring, among novel things, to the Chechinquamin (chinkapin), and another which no one can fail to recognize.

Plums, [he says], are of three sorts. . . . That which they call *Putchamins* grow as high as a *Palmeta;* the fruit is like a Medler; it is first greene, then yellow, and red when it is ripe; if it be not ripe it will draw a man's mouth awry with much torment.[32]

In his description of New England, Smith mentions twelve species of mammals, including the "moos," now spoken of for the first time,[33] sixteen of birds, and twenty-seven "fishes." His descriptions of the abundance of fishes are often quoted.[34]

Smith's first work upon Virginia was printed in 1612 and his General History in 1624. In the interim, Ralphe Hamor, the younger, secretary of the colony, issued his True Discourse of the Present Estate of Virginia, published in London in 1615.[35] Hamor was not a naturalist, but his name is usually referred to by zoological bibliographers, since he mentions by name over sixty native animals. He was the first to describe the great flocks of wild pigeons, of which he remarks: "In winter, beyond number or imagination, myselfe hath seene three or foure houres together flockes in the aire so thicke that even they have shadowed the skie from us."[36] He gives an amusing description of the "opossume," and also speaks of the introduction and successful acclimation of the Chinese silkworm.

In 1620 the Plymouth Colony was planted, and its members also began to record their impressions of the birds and the beasts and the plants which they found, for the instruction of their kinsfolk at home.

Bradford and Winslow's Journal, printed in London in 1622, contains various passing allusions to the animals and plants observed by the Pilgrims, as does also Bradford's History, which, however, was not printed until long after its completion. They added nothing, however, to what had already been said by Smith.

Edward Winslow's News from New England, printed in London in 1624, contains one of the earliest descriptions of the Indians of the Northeast.

William Wood's New England's Prospect, which was issued in London in 1634, and Morton's New English Canaan, printed three years later in Amsterdam, were the first formal treatises upon New England and its animals and plants. The two authors were very unlike, and their books even more so—yet complementing each other very satisfactorily. Morton was the best educated man, brightest, and most observant; Wood the most conscientious and the most laborious in recording minute details.

"Thomas Morton, of Clifford's Inn, Gent.," was by no means a representative man in the Puritan community in which he lived. His habits were those of an English man of fashion, and his Rabelaisian humor, when directed against his fellow-colonists and their institutions, was no recommendation to their favor. We can not wonder that he was hunted from settlement to settlement and even cast into prison, to endure, without bedding or fire, the rigor of a New England winter.

As a naturalist, Morton appears to have been the most accurate of the two of this time. In those parts of his book which describe animals and plants he manifests a definite scientific purpose. He discriminates between species, and frequently points out characters by which American and European forms may be distinguished. He was the first to banish the lion from the catalogue of the mammals of eastern North America. Even Wood, though he admitted that he could not say that he ever saw one with his own eye, evidently believed that lions inhabited the woods of Masschusetts. Morton was a skeptic because, as he said, "it is contrary to the Nature of the beast to frequent places accustomed to snow; being like the Catt, that will hazard the burning of her tayle, rather than abide from the fire." His brief biographies, especially those of mammals, indicate that he was an observer of no slight acuteness.

Twenty species of mammals, thirty-two of birds, twenty of fishes, eight of marine invertebrates, and twenty-seven of plants are mentioned, usually in such definite terms that they may readily be identified.

A thorough pagan himself, he seems to have commanded the confidence of the Indians more than others, to have lived in their society, and learned to comprehend the meaning of their customs. His first book, The Originall of the Natives, their Manners and Customs, seems to have been the careful record of rather critical observations.

Wood's book is no less deserving of praise. The climate and the soil are judiciously discussed, and the herbs, fruits, woods, waters, and minerals, then "the beasts that live on land," "beasts living in the water," "birds and fowls both of land and water," and fish, after which follows a topographical description of the colony. His catalogues of species are in verse, and his adjectives are so descriptive and pictorial that his subsequent remarks in prose are often superfluous. I quote his catalogue of the trees of New England, an imitation in manner and meter of Spenser's famous catalogue in The Faerie Queene:

> Trees both in hills and plaines in plenty be
> The long liv'd Oake, and mourneful Cypris tree
> Skie towring pines, and Chestnuts coated rough,
> The lasting Cedar and the Walnut tough;
> The rozin dropping Firre for masts in use.
> The boatmen seeke for Oares light neeate growne sprewse,
> The brittle Ash, the ever trembling Aspes,
> The broad-spread Elme, whose concave harbours waspes
> The water-springie Alder, good for nought
> Small Elderes by the Indian Fletchers sought
> The knottie Maple, pallid Birtch, Hawthornes,
> The Horne bound tree that to be cloven scornes;
> Which from the tender Vine oft takes his spouse,
> Who twinds embracing armes about his boughes.
> Within this Indian orchard fruites be some
> The ruddie Cherrie, and the jettie Plumbe
> Snake murthering Hazell, with sweet Saxaphrage
> Whose steemes in beere allays hot fever's rage.
> The Diar's Shumach, with more trees there be
> That are both good to use and rare to see.

Thus he describes the Animals of New England:

> The Kingly Lyon and the strong arm'd Beare
> The large limbed Mooses, with the tripping Deare.
> Quill darting Porcupines, and Rackcoones bee
> Castelld in the hollow of an aged Tree
> The skipping Squirrel, Rabbet, purblinde Hare
> Immured in the selfe same Castle are
> Least red-eyed Ferrets, wily Foxes should

Them undermine if ramperd but with mould.
The grim fac't Ounce, and ravenous howling Woolfe
Whose meagre Paunch suckes like a swallowing Gulfe,
Black glistening Otters and rich coated Beaver
The Civet scented Musquash, smelling ever.

His subsequent remarks upon the mammals are expanded from his rhyme, and extended by tales which he has heard from hunters. One of the animals whose name would not lend itself to poesy is the "squuncke," which he classified among the "beasts of offence." This seems to be the first use of the name.

In the second part of Wood's book the Indians are discussed, and a very creditable vocabulary is given.

Most admirable work was now being done among the Indians by some of the colonial clergymen. Chief among them was the Rev. John Eliot [b. 1604, d. 1690], who, during a residence of more than half a century at Roxbury, mastered the language of the Massachusetts branch of the great Algonquin tribe and published his grammars and translations. He was a graduate of Jesus College, Cambridge, and came to Massachusetts in 1631. The Rev. Abraham Peirson, one of the founders of the colony at Newark, during his residence in New England made valuable investigations upon the language of the Quiripi or Quinnipiac Indians of the New Haven colony. The extensive bibliography of which Mr. Pilling has recently published advance sheets gives an excellent idea of the attention which American linguistics have since received.

That very eminent colonial stateman, John Winthrop the younger, the first governor of Connecticut [b. 1587, d. 1649], stood high in the esteem of English men of science, and was invited by the newly founded Royal Society, of which he was a fellow, "to take upon himself the charge of being the chief correspondent in the West, as Sir Philiberto Vernatti was in the East Indies." The secretary of the Royal Society said of him: "His name, had he put it to his writings, would have been as universally known as the Boyles, and Wilkins's, and the Oldenburghs, and been handed down to us with similar applause." [37]

Governor Winthrop's name occurs from time to time in the Philosophical Transactions, and it was to him that science was indebted for its first knowledge of the genus *Astrophyton*.

John Winthrop, F. R. S. [b. 1606, d. 1676], son of the last, and also governor of Connecticut in 1662, is said to have been "famous for his philosophical knowledge." He was a founder of the Royal Society, being at the time of its origin in England as agent of the colony. And the second governor's grandson, John Winthrop, F. R. S. [b. 1681, d. 1747], who passed the latter part of his life in England, was declared to have increased the Royal Society's repository "with more than six hundred curious specimens, chiefly in the mineral kingdom," and since the founder of the museum of the Royal Society. "the benefactor who has given the most numerous collections." [38]

The Rev. John Clayton, rector of Crofton, at Wakefield, in Yorkshire, made a journey to Virginia in 1685, and in 1688 communicated to the Royal Society An Account of several observables in Virginia and in his Voyage thither. [39] Clayton seems to have been a man of scientific culture, and to have been the author, in company with Doctor Moulin, of a treatise upon comparative anatomy. He was of the same school with Harriot and Wood, though more philosophical. His essay was, however, the most important which had yet been published upon the natural history of the South, and his annotated catalogue of mammals, birds, and reptiles is creditably full.

Thomas Glover also published about this time An Account of Virginia, [40] in which he discussed the natural history of the colony after the manner of Wood and Morton. The Rev. Hugh Jones also published a similar but shorter paper upon Several Observables in Maryland, [41] in which, however, no new facts are mentioned. He collected insects and plants for Petiver.

Benjamin Bullivant, of Boston, was another of the men who, to use the language of the day, was "curious" in manners of natural history. One of his letters was published in the Philosophical Transactions,[42] and his notes on the "hum-bird" are sometimes referred to.

Bullivant was not a naturalist; he is less worthy of our consideration than Harriot, although a century later. A fit companion for Bullivant was John Josselyn.

Josselyn's famous work entitled New England's Rarities Discovered in Birds, Beasts, Fishes, Serpents, and Plants of that Country, was printed in London in 1672; his Account of Two Voyages to New England, in 1675 (second edition). No writer of his period is more frequently quoted than Josselyn, whose quaint language and picturesque style are very attractive. Although no more in sympathy with his Puritan associations than the author of New England's Prospect, he was evidently more justly entitled to subscribe himself as "Gentleman," and his books are not disfigured by personalities and political aspersions.

Josselyn does not seem to me to be the peer, as a naturalist, of many of those who preceded him. He was a bright, though superficial, man, and a ready compiler. He evidently had some botanical work in his possession, possibly, as Tuckerman has suggested, a recently published edition of Gerard's Herbal, and this he used with such skill as to give him a certain standing in botanical literature. In his zoological chapters I find little which had not been recorded before, while the author's fondness for startling anecdotes greatly mars the semblance of accuracy in his work. His catalogue of fishes is a strange olla-podrida of names and scraps of information, compiled, collected, and invented. His method of arrangement is not more scientific than his spirit, and it is questionable whether he is entitled to a place among naturalists.

Here is an example of his style:

"The *Basse*," writes he, "is a salt water fish too . . . . one writes that the fat in the bone of a *Basses* head is his braines which is a lye."

To this period belongs, also, Lawson, the author of a History of Carolina and A New Voyage to Carolina, made in 1700 and the following years, while acting as surveyor-general of the colony. Lawson was burnt at the stake in 1709 by the Indians, who resented his encroachments upon their territory. His lists of the animals and plants of the region are very full and his observations accurate. Coues's "Lawsonian period" in the history of American ornithology is hardly justifiable. Lawson belonged to the school of Harriot and the first Clayton.

Edward Bohun and Job Lord, of Carolina, appear to have been interested in natural history at this time and to have been collecting specimens for Petiver in London, while William Vernon was engaged in similar occupations in Maryland.

In those early days all Europe was anxious to hear of the wonders of America, and still more eager to see the strange objects which explorers might be able to preserve and bring back with them. Public museums were as yet unknown, but the reigning princes sought eagerly to secure novelties in the shape of animals and plants.

Columbus was charged by Queen Isabella to collect birds, and it is recorded that he took back to Spain various skins of beasts. Even to this day may be seen, in Siena, hanging over the walls of the old collegiate church, a votive offering, placed there nearly four centuries ago by the discoverer of America, then in the prime of his glory. It consists of the helmet and armor worn by him when he first stepped upon the soil of the New World, and the rostrum of a swordfish killed on the American coast.

The State papers of Great Britain contain many entries of interest to naturalists. King James I was an enthusiastic collector. December 15, 1609, Lord Southampton wrote to Lord Salisbury that he had told the King of the Virginia squirrels brought into England, which were said to fly. The King very earnestly asked if none were provided for him— whether Salisbury had none for him—and said he was sure Salisbury would get him one. The writer apologizes for troubling Lord Salisbury, "but," he continues, "you know so well how he (the King) is affected to such toys."

Charles I appears to have been equally curious in such matters. In 1637 he sent John Tradescant, the younger, to Virginia "to gather all rarities of flowers, plants, and shells."

In 1625 we find Tradescant writing to one Nicholas that it is the Duke of Buckingham's pleasure that he should deal with all merchants from all places, but especially from Virginia, Bermudas, Newfoundland, Guinea, the Amazons, and the East Indies for all manner of rare beasts, fowls and birds, shells and shining stones, etc.[43]

In the Domestic Correspondence of Charles I, in another place,[44] July, 1625, is a "Note of things desired from Guinea, for which letters are to be written to the merchants of the Guinea Company." Among other items referred to are "an elephant's head with the teeth very large; a river horse's head; strange sorts of fowls; birds and fishes' skins; great flying and sucking fishes; all sorts of serpents, dried fruits, shining stones, etc." Still further on is a note of one Jeremy Blackman's charge, in all £20, for transporting four deer from Virginia, including corn and a place made of wood for them to lie in.[45]

Not only did the kings make collections, but the keepers of public houses made museums then, as they do now, for the pleasure of their patrons.

At the middle of the last century there appear to have been several collections of curiosities.

In Artedi's ichthyological works there are numerous references to places where he had seen American fishes, especially at Spring-garden[46] and at the Naggshead, and the White-bear, and the Green Dragon in Stepney, in those days a famous hostelry in London. He speaks also of collections at the houses of Mr. Lillia and Master Saltero's[47] in Chelsey and at Stratford, and also in the collection of Seba, in Amsterdam, and in that of Hans Sloane.

With the exception of "*the monk* or *Angel-fish, Anglis* aliis *Mermaid-fish*," probably a species of *Squatina*, which he saw at the Nag's Head, all the fishes in these London collections belonged to the order Plectognathi.

Josselyn, after telling us how a Piscataway colonist had the fortune to kill a Pilhannaw—the king of birds of prey—continues, "How he disposed of her I know not, but had he taken her alive and sent her over into England neither Bartholomew nor Sturbridge Fair could have produced such another sight." [48]

Shakespeare's mirror strongly reflects the spirit of the day. When Trinculo, cast ashore upon a lonesome island, catches a glimpse of Caliban he exclaims:

"What have we here,—a man or a fish? Dead or alive? A fish: he smells like a fish; a very ancient and fish-like smell. . . . A strange fish! Were I in England now, (as once I was,) and had but this fish painted, not a holiday fool there but would give a piece of silver; there would this monster make a man; any strange beast there makes a man: when they will not give a doit to relieve a lame beggar, they will lay out ten to see a dead Indian." [49]

The compilers of the great encyclopedialike works on natural history were quick to pick up the names and descriptions of the American animals which had found their way to Europe, and many such are mentioned in the writings of Gesner, Clusius and Aldrovandus, Lister, Laet, and Willughby. [50]

Creatures of remarkable appearance, which could be preserved with ease, were the first to become known. Among fishes, for instance, those with a hard, inflexible integument, such as the trunk fishes. Every species of the family *Ostraciontidæ* was known in Europe as early as 1685; most of them probably a century before. We know that Columbus caught a trunk fish and described it in his Voyages.

Professor Tuckerman has traced in a most instructive manner the beginnings of European acquaintance with American plants, finding traces of the knowledge of a few at a very early period:

Dalechamp, Clusius, Lobel, and Alpinus—all authors of the sixteenth century—must be cited occasionally in any complete synonymy of our *Flora*. The Indian-corn, the side-saddle flower (*Sarracenia purpurea* and *S. flava*), the columbine, the common milkweed (*Asclepias cornuti*), the everlasting (*Antennaria margaritacea*), and the *Arbor vitæ*, were known to the just-mentioned botanists before 1600. *Sarracenia flava* was sent either from Virginia, or possibly from some Spanish monk in Florida. Clusius's figure of our well-known northern *S. purpurea* . . . . was derived from a specimen furnished to him by one Mr. Claude Gonier, apothecary at Paris, who himself had it from Lisbon; whither we may suppose it was carried by some fisherman from the Newfoundland coast. The evening primrose (*Œnothera biennis*) was known in Europe, according to Linnæus, as early as 1614. *Polygonum sagittatum* and *arafolium* (tear-thumb) were figured by De Laet, probably from New York specimens, in his Novus Orbis, 1633. Johnson's edition of Gerard's Herbal (1636). . . . contains some dozen North American species, furnished often from the garden of Mr. John Tradescant . . . . and John Parkinson—whose Theatrum Botanicum (1640) is declared by Tournefort to embrace a larger number of species than any work which had gone before it—describes, especially from Cornuti, a still larger number.[51]

All the early voyagers were striving for the discovery of a western passage to India, and the West Indies, so called, were considered simply a stage on the journey toward the East Indies. It is not strange, therefore, that writers should often have failed to distinguish the faunal relations of the animals which they described. Many curious paradoxes in nomenclature have thus arisen—*Cassis madagascariensis*, for instance, a very misleading name for a common West India mollusk.

V

The seventeenth century bears upon its roll the names of many explorers besides those of English origin who have already been named. Within fifty years of the time of Harriot and of the planting of the colony at Roanoke, the number and extent of the European settlements in America had become very considerable. Virginia and the New England plantations were growing populous and Maryland was fairly established. Insular colonies were thriving at Newfoundland and Bermuda and on Barbados and elsewhere in the West Indies.

New Spain and Florida marked the northern limits of the domain of the Spaniards, who had already overrun almost all of South America.

New France bounded New England on the north, and the French were pushing their military posts and missionary stations down into the Mississippi Valley.

The Dutch were established on Manhattan Island and elsewhere in the surrounding country, and the Dutch West India Company had already a foothold in Brazil and Guiana. A colony of Scandinavians had been planted by the Swedish West India Company near the present site of Philadelphia, and the forsaken Danish colonies of Greenland were soon to be reestablished. The Portuguese had flourishing settlements in Brazil, for the possession of which they were contending with the Dutch.

Every European nation was represented in the great struggle for territory save Italy and Germany, Switzerland and Russia; but the Italians and Germans, the Swiss and the Russians were to hold their own in the more generous emulation of scientific exploration which was to follow.

During the seventeenth and eighteenth centuries numerous explorations were made both in North and South America by Spanish, French, Dutch, German, and Scandinavian explorers. Although these men have been studied in the preparation of this address, I do not intend to speak of them at any length, but to confine my attention in the main to the growth of scientific opinions and institutions in the English colonies.

The number of volumes of reports and narratives, often sumptuously printed and expensively illustrated, which were published during the seventeenth and eighteenth centuries, impresses upon one most powerfully the idea of the earnestness, diligence, and intelligence of their writers.

The Spaniards.——Even as early as the beginning of the century, Spanish influence was less prominent in the affairs of the New World; in no respect more strikingly so than in explorations. The political supremacy of Spain was gone, her intellectual activity was waning, and the mighty storm of energy, by which her domain in America had been so suddenly and widely established, seemed to have completely exhausted the energy of her people, depleted as it had been by wars without and religious prosecution within.

From this time forward the record of Spanish achievements in the fields of science and discovery is very meager. Between the day of Hernandez and that of Azara and Mutiz, who explored South America in the latter part of the eighteenth century, I find but two names worthy of mention, and these seem properly to belong with the naturalists who lived a hundred years before them. I refer to José Gumilla, who published, in 1741, a work on the natural history of the Orinoco region, and Miguèl Venegas, whose Noticia de la California appeared in 1757.

The French.——One of the first French explorers who left a record of his observations was Samuel de Champlain, who made a voyage to the West Indies and Mexico, 1599–1602, and began his travels in New France in 1603. He was the founder of Quebec, where he died in 1635, and his geographical explorations and maps are of great value. His observations upon the animals and plants are disappointing. He describes the gar-pike and the king-crab, already described and figured by Harriot many years before, and refers in unmistakable terms to the shearwater, the caribou, the wild turkey, and the scarlet tanager. His lists of animals which occur now and again in the course of his narrative are too vague to be of value.[52]

Much higher in the esteem of naturalists was Gabriel Sagard Théodat, a Franciscan friar, whose Le Grand Voyage du Pays des Hurons, printed in 1632, was the most scholarly work upon America which had yet appeared, and whose History of Canada and the journeys made by the Franciscans for the conversion of the infidels also contains most valuable records.

The first work on the plants of North America was that of Cornuti— Canadensium Plantarum, aliarùmque nondum editarum historia— printed in Paris in 1635, which described thirty-seven species, thirty-six of these being illustrated by elaborate engravings upon copper. The botanical part of his treatise is usually ascribed to Vespasian Robin, and Tuckerman supposes that the local notes, as well as specimens described, were probably the result of the labors of the worthy Franciscan missionary, Sagard.[53]

A few years later, Pierre Francois Xavier de Charlevoix [b. 1682, d. 1761], a Jesuit priest, having by royal command traveled through the northern part of North America, published his Histoire et Description Générale de la Nouvelle France, Paris, 1744, which was full of important biological and ethnological observations, the accuracy of which is not questioned.

He subsequently traveled in South America, and published in 1760 a work full of statements concerning the animals, plants, and fruits of that country, and also particularly interesting from the account of which it gives of the singular Jesuit establishment in Paraguay.

Other French missionaries, Breboeuf, Du Poisson, Jaques, Joliet, La Chaise, Lallemand, Marquette, Senat, and Souel, followed Charlevoix in the exploration of these regions. Their works contain many valuable notes upon animals and plants.

Jean Baptiste du Tertre, in his Histoire Générale des Antilles, habitées par les François, published in Paris in 1667 [ed. 1667–71], described and illustrated many of the New World animals.

In 1672 Nicolas Denyse published in Paris two comprehensive works upon America, viz: Histoire Naturelle des Peuples, des Animaux des Arbres and Plantes de l'Amérique,[54] and Description Geographique des Costes de l'Amérique Septentrionale, avec l'Histoire Naturelle du Païs.[55]

F. Froger, a companion of De Gennes in his voyage made in 1695–1697 to the coast of Africa, the Straits of Magellan, Brazil, Cayenne, and the Antilles, published a report in 1698.[56] The book has been overlooked by recent bibliographers, but, judging from Artedi's remarks upon its ichthyological portion, it was fully equal to similar works of its day.

Baron de la Hontan, lord lieutenant of the French colony at Placentia, printed at the Hague in 1703 his Voyages dans l'Amérique, which is sometimes referred to by zoologists.

Louis Feuillée, who traveled by royal commission from 1707 to 1712 in Central and South America, published four volumes of physical mathematics and botanical observations, 1714–1725, in Paris.

The Père Jean Baptiste Labat visited the West Indies as a missionary early in the eighteenth century, and Nouveau Voyage aux Isles de l'Amérique, printed in Paris, 1722, is very full of interesting and copious details of natural history.

The Père Laval visited Louisiana and published in Paris, 1728, his Voyage de la Louisiane.

M. Le Page Du Pratz followed, in 1758, with his Histoire de la Louisiane,[57] full of geographical, biological and anthropological observations upon the lower valley of the Mississippi, and Captain Bossu, of the French marines, also published upon the same region,[58] translated into English in 1771 by John Reinhold Forster, whose notes gave to the work its only value. These men are all catalogued with the seventeenth-century naturalists because they were of the old school of general observers and only indirectly contributed to the progress of systematic zoology.

Charles Plumier [b. 1646, d. 1704] was sent thrice by the King of France to the Antilles during the latter years of the seventeenth century. He published three magnificently illustrated works upon the plants of America[59] and left an extensive collection of notes and drawings of animals and plants, many of which have proved of value to naturalists of recent years. His colored drawings of fishes were of great service to Cuvier in the preparation of his great work upon ichthyology, and in some instances species were founded upon them.

The Dutch.—There were few lovers of nature among the colonists of Manhattan, and with the exception of certain names which have clung to well-known animals, such as the mossbunker and weakfish, naturalists have little to remind them of the days of Van Twiller and Stuyvesant. Van Der Donck, in 1659, described the fauna, and Jakob Steendam's poem, "In praise of the Netherlands," catalogued many of the animals.

The achievements of Prince Maurice of Nassau (b. 1604, d. 1679), the conqueror of Brazil, during his residence in that country from 1636 to 1644, were far more important than those of any one man in the seventeenth century, and entitled the Netherlands to a leading place in the early history of American scientific explorations. The notes and figures which were collected by him and his scientific assistants, Marcgrave, Piso, and Cralitz, were published in part under the editorship of Golius and Laet, and have been frequently used by naturalists of the present century. An atlas of colored drawings from the hand of Prince Maurice is still preserved in the Royal Library in Berlin. Here are depicted 34 species of mammals, 100 of birds, 55 of reptiles, 69 of fishes, and 77 of insects, besides many of plants.

Marcgrave's's Historia Rerum Naturalium Brasiliæ was printed in Amsterdam in 1648, four years after his untimely death while exploring the coast of Guinea.

Piso's Medicina Braziliensis, 1648, and his Natural History and Medicine of both Indies, 1658, were also results of Prince Maurice's expedition.

Among other contributions made by the Netherlands to the natural history of America were the Relation de Voyage de Isle Tobago, Paris, 1606, and the Histoire Naturelle et Morale des Iles Antilles, Rotterdam, 1658,[60] written by N. Rochefort, a Protestant missionary to the West Indies, and Jan Nieuhof's See und Landreize benessens een bondege Beschreyving van gantsch Nederland Brazil so van Landschappen Steden, deren Gewaffen, etc., printed in 1682.

Jan Jacob Hartsinck, a Dutch traveler in Guiana, printed a book of scientific travels at Amsterdam in 1770.

Philippe Fermin, a Dutch naturalist, a resident for many years in Surinam, published in Amsterdam two important works upon the natural history of that region, in 1765 his Histoire Naturelle de la Hollande Equinoxiale, and in 1769 his Description de Surinam. I refer to these works as important, not because they are of great value to zoological writers of to-day, but because they, in their day, marked distinct advances in knowledge.

The Scandinavians.—Danish enterprise at an early day sent explorers to the Western Continent, and the scholarly tendencies of the Scandinavian mind were soon manifest in a literature of geographical and scientific observations.

Hans Egede, a missionary who went to Greenland at least as early as 1715, published in 1741 his comprehensive work upon Greenland, of which so many editions have been published.

Otho Fabricius [b. 1744, d. 1822], another missionary, long resident in Greenland, published in 1780 his Fauna Grœnlandica, a work which in scientific accuracy has never been excelled—a most important contribution to systematic zoology. David Crantz's History of Greenland, published in 1770, is another important scientific work from the hand of a missionary, and Zorgdrager's notices of the Greenland fisheries deserve a passing notice.

The travels of Kalm, a Swede and a pupil of Linnæus, are noticed elsewhere. Peter Loefling, another pupil of Linnæus, visited Spanish America, and in his Iter Hispanicum, printed in Stockholm, 1758, described many animals and plants observed by him.

Olaf Swartz, a Swede, discovered and described 850 new species of West Indian plants from 1785 to 1789. He spent a year in the Southern United States before going to the West Indies.[61]

The Germans.——Germany, too, soon began to send its students across the Atlantic. Johann Anderson, a burgomaster of Hamburg, published in 1746 his Tidings from Iceland, Greenland, and Davis Straits, for the benefit of Science and Commerce. Hans Just Winkelmann published in Oldenburg in 1664 Der Amerikanischen neuen Welt Bescreibung, etc., with descriptions and figures of animals and plants.

Christian Bullen in 1667 made a voyage to Greenland and Spitzbergen, an account of which, including interesting observations on whales and the whale fishery, was printed at Bremen in 1668.

Marcgrave, Krieg, the two Forsters, and Schoepf are referred to elsewhere. Steller, Pallas, and Chamisso are mentioned in connection with Russian explorations.

Madame Maria Sibilla Merian [b. 1647, d. 1717], who was a native of Frankfort, was an enthusiastic entomologist who traveled in Surinam from 1699 to 1701. Her paintings of tropical insects were reproduced in a magnificent folio volume, printed 1705–1709, which was one of the wonders of her day, and which, together with her other writings upon insects, have secured her a prominent place in the early history of science.

V I

The seventeenth century was not, upon the whole, a period favorable to the promotion of science, for all Europe was agitated by war and political strife, and men had neither opportunity nor inclination for intellectual pursuits. During its latter half, however, and with the return of peace and tranquility, science grew in favor as it had never done before. The restoration of the Stuarts to the English throne was quickly followed by the establishment of the Royal Society. Louis XIV made the period of his accession memorable by founding the Royal Academy of Sciences, and by building an observatory.

This was the period of intellectual activity which followed the revival of letters in Europe. Carus, in his Geschichte der Zoologie, 1872, p. 259, calls it the period of encyclopædia-making (Periode der encyklopädischen Darstellungen), filling the interspace between "The Zoology of the Middle Ages" and "the period of Systematic Classification." Students of science had ceased to compile endless commentaries on the works of Aristotle, and had begin to record their own observations and thoughts, to gather new facts and materials, which were to serve as a basis for the systematic work for their successors.

The greatest names of the day among naturalists were those of Ray, Tournefort, Lister, Jonston, Goedart, Redi, Willughby, Swammerdam, Sloane, Jung, and Morrison; names not often referred to at the present day, but worthy of our recollection and veneration, for they were men of a new era—the pioneers in systematic zoology and botany.

Among the earliest representatives of the new school in North America were Banister, Clayton, Mitchell, and Garden. John Banister, a clergyman of the Church of England, emigrated to Virginia before 1668, and in addition to his clerical duties applied himself assiduously to the study of natural history. He was a disciple and also, no doubt, a pupil of the great English naturalist, John Ray, who called him in his Historia Plantarum, "erudissimus vir et consummatissimus Botanicus," and corresponded also with Lister, and Compton, Bishop of London. He was the first to observe intelligently the mollusks and insects of North America. In a paper communicated to the Royal Society in 1693 he refers to drawings of ten or twelve kinds of land snails and six of fresh-water mussels. The drawings were not published, nor were the notes, except those in reference to the circulation of a species of snail. [62]

He sent to Petiver, in 1680, a collection of fifty-two species of insects, his observations upon which, with notes by Petiver, were a few years later communicated to the Royal Society.[63] Among them many familiar forms are recognizable—the mudwasp, seventeen-year locust, cimex, cockroach, firefly, the spring beetle (*Elater*), and the tobacco moth. He appears to have drawn and described several phases of the life history of the ichneumon fly. He had in his possession in 1686, and exhibited to an English traveler, large bones and teeth of fossil mammals from the interior of Virginia, the first of which we have any record in North America.[64]

It was as a botanist, however, that he was best known. He made drawings of the rarer species, and transmitted these with his notes and dried specimens to Compton and Ray. Banister's Catalogus Plantarum in Virginia Observatarum, printed in 1686,[65] was the first systematic paper upon natural history which emanated from America. In one of his botanical excursions, about the year 1692, he visited the falls of the Roanoke, and, slipping among the rocks, was killed.[66]

Lawson, the historian of North Carolina, writing at the beginning of the next century, remarked: "Had not the ingenious Mr. Banister (the greatest virtuoso we ever had on this continent) been unfortunately taken out of this world, he would have given the best account of the plants of America of any that ever yet made such an attempt in these parts."[67] The memory of John Banister is still cherished in Virginia, where his descendants are numerous.[68]

John Clayton was also an excellent representative of the new school, and should not be confounded with the Rev. John Clayton who visited America in 1685. John Clayton, the naturalist, as he is styled in Virginian history, appears to have been born in Fulham, a suburb of London, in 1693, and to have accompanied his father, John Clayton, subsequently attorney-general of Virginia, when he came to this country in 1705. He was clerk of Gloucester County, Virginia, for fifty-one years, and died December 15, 1773. "He passed a long life," says Thacher, "in exploring and describing the plants of this country, and is supposed to have enlarged the botanical catalogue as much as any man who ever lived." He was a correspondent of Linnæus, Gronovius, and other naturalists, as well as of Collinson, who wrote of him in 1764 as "my friend John Clayton, the greatest botanist of America."

Clayton's Flora Virginica, which was edited by J. F. Gronovius, assisted by the young Linnæus, who was just entering upon his career of success and was then resident in Leyden, began to appear in 1739, subsequent portions being published in 1743 and 1762. It seems to be the opinion of botanists that Gronovius deserves less credit for his share in this work than has usually been allowed him, and that Clayton's descriptions were those of a thorough master of botanical science as then understood. He communicated to the Royal Society various botanical papers, including one upon the culture of the different kinds of tobacco. On his death he left two volumes of manuscripts, and an herbarium, with marginal notes and references for the engraver who should prepare the plates for his proposed work. These were in the possession of his son when the Revolutionary war commenced, and were placed in the office of the clerk of New Kent County for security from the invading enemy. The building was burned down by incendiaries, and thus perished not only the records of the county, but probably one of the most important works on American botany written before the days of Gray and Torrey.

Jefferson declares that Clayton was a native Virginian, and such is the confusion in the records that it is quite possible that such may be the fact.[69]

Still another pioneer was Doctor John Mitchell, born in England about 1680, and settled early in the last century at Urbana, Virginia, on the Rappahannock, where he remained nearly fifty years, practicing medicine and promoting science. He appears to have been a man of genius and broad culture, and was one of the earliest chemists and physicists in America. His political and botanical writings were well received, and his map of North America is still an authority in boundary matters. He was a correspondent of Linnæus, and in 1740 sent Collinson a paper in which thirty new genera of Virginia plants were proposed.[70] His Dissertation upon the Elements of Botany and Zoology[71] was dated Virginia, 1738, and was thus almost contemporary with the first edition of the Systema Naturæ of Linnæus, though it was not printed until ten years after it was written. This was the first work upon the principles of science ever written in America. In 1743 he communicated to the Royal Society An Essay upon the Causes of the different Colours of People in different Climates, writing[72] from the standpoint of an evolutionist. He also communicated An Account of the Preparation and Uses of the various Kinds of Potash,[73] and a letter concerning the Force of electrical cohesion.[74] His fame rests chiefly, however, upon his investigations into the yellow fever epidemic of 1737–1742, published after his death by his friends, Franklin and Rush.[75] In 1743 he appears to have been engaged in physiological researches upon the opossum, which, however, were never published. In 1746 Doctor Mitchell returned to England, and upon the voyage was captured by French or Spanish pirates, and his collections and apparently his manuscripts destroyed. He became a Fellow of the Royal Society, and in 1748 was writing a work upon the natural and medical history of North America.[76] He died at an advanced age, about 1772. His name is perpetuated in that of our beautiful little partridge berry, *Mitchella repens*. "Mitchell and Clayton together, "says Tuckerman, "gave to the botany of Virginia a distinguished luster."

Doctor John Tennent, of Port Royal, Virginia, seems to have been a man of botanical tastes. He it was who brought into view the virtues of the Seneca snake root, publishing at Williamsburg, in 1736, an essay on pleurisy, in which he treats of the Seneca as an efficient remedy in the cure of this disease.[77] He also wrote other botanical treatises.[78] Doctor George Greham, of Dumfries, Virginia, was a man of similar tastes, and it is said by Mr. Jefferson that we are indebted to him for the introduction to America of the tomato.

David Krieg, F. R. S., a German botanist, collected insects for Petiver in Maryland, and gathered also hundreds of species of plants. He seems to have returned to England very early in the century, for his name appears in the Philosophical Transactions in 1701.

Colonel William Byrd, of Westover, Virginia, [b. 1764, d. 1793], was a man of European education, the owner of a magnificent library, in which Stith wrote his history of Virginia, founder of the city of Richmond, colonial agent in London, and president of the King's council. He was a fellow of the Royal Society, to which he communicated a paper An Account of a Negro Boy that is dappeld in several Places of his body with White spots,[79] and was a correspondent of Collinson, Bartram, and other naturalists. His History of the Dividing Line, and his Journey to the Land of Eden, in 1733, contain many interesting observations upon Indians and general natural history. He it was who, in 1694 carried to England a female opossum, which furnished the materials for the first dissertation upon the anatomy of the marsupiates. [80]

One of the most eminent of our colonial naturalists was Doctor Alexander Garden, born in Scotland about 1728 [d. 1791]. He emigrated to America about 1750, and practiced medicine in Charleston, South Carolina, until after the close of the Revolutionary war, when he returned to England and became very prominent in scientific and literary circles, and vice-president of the Royal Society in 1783. He was an excellent botanist, but he did his best work upon fishes and reptiles. He sent large collections of fishes to Linnæus, which were so well prepared that when I examined the fishes in the Linnæan collection in London, in 1883, I found nearly every specimen referred to by him in his letters in excellent condition, though few collected by others were identifiable. Garden was the discoverer of *Amphiuma means*, and was instrumental in first sending the electrical eel to Europe. His letters to Linnæus and to Ellis are voluminous and abound in valuable information. In 1764 he published a description of *Spigelia marilandica*, with an account of its medicinal properties.

James Logan [b. 1664, d. 1751], a native of Ireland and member of the Society of Friends, accompanied William Penn to this country in 1682 in the capacity of secretary, and became a public man of prominence, serving for two years as governor of the colony of Pennsylvania. He was a man of broad culture and was the author of a translation of Cicero's De Senectute, printed by Benjamin Franklin in 1744. To Logan belongs the honor of having carried on the first American investigations in physiological botany, the results of which were published in Leyden, in 1739, in an essay entitled Experimenta et Meletemata de Plantarum Generationis. This essay, which related to the fructification of the Indian corn, was accepted in its day as a valuable contribution to knowledge.

Cadwallader Colden [b. 1688, d. 1776] was also a statesman and a naturalist. A native of Scotland, he came to America in 1708, and, after a short residence in Pennsylvania, settled in New York, where he held the office of surveyor-general and member of the King's council, and in later life was for many years lieutenant-governor, and frequently acting governor of the province. His intellectual activity manifested itself in various directions, and his History of the Five Indian Nations of Canada, New York, 1727, was one of the earliest ethnological works printed in America. He was also interested in meteorology and astronomy, and as a correspondent of Linnæus and Collinson did much to advance the study of American botany. His daughter, Miss Jane Colden, was the first lady in America to become proficient in the study of plants. She was the author of a Flora of New York, which was never published.[81] Governor Colden's Plantæ Coldenhamiæ, the first part of a catalogue of the plants growing in the neighborhood of his country residence, Coldenham, near Newburg, was the first treatise on the flora of New York. It was published in 1744 in the acts of the Royal Society of Upsala.[82] A most interesting collection of papers from the scientific correspondence of Colden was published many years ago by Doctor Asa Gray.[83]

Hans Sloane, a young Irish physician [b. 1660, d. 1753], who had been a pupil of Tournefort and Magnol, visited the West Indies in 1684, and after his return printed a Catalogue of Jamaica Plants in 1696, and later a sumptuously illustrated work on the natural history of Jamaica (1707–1725). After his return he became an eminent physician, and in 1727 succeeded Isaac Newton as president of the Royal Society. The collection of animals and plants made by Sir Hans Sloane in America was greatly increased by him during his long and active life, and, having been bequeathed by him to the nation, became, upon his death in 1753, the nucleus of the British Museum.

Another naturalist of the same general character was Mark Catesby [b. 1679, d. 1749], who lived in Virginia, 1712 to 1721, collecting and making paintings of birds and plants; in the Carolinas, 1722 to 1725, and a year also in the Bahamas. His magnificent illustrated work upon the Natural History of Carolina, Florida, and the Bahama Islands,[84] is still of great value to students of natural history.

The name of John Bartram, the Quaker naturalist of Philadelphia, is possibly better remembered than those of his contemporaries. This is no doubt due to the fact that he left behind him a lasting monument in his botanic garden on the banks of the Schuylkill. He was the earliest native American to prosecute studies in systematic botany, unless Jefferson's statement concerning Clayton proves to be true. Linnæus is said to have called him "the greatest natural botanist in the world," and George III honored him in 1765 with the title of Botanist to his Majesty for the Floridas and a pension of £50 a year. Bartram was a most picturesque and interesting personage, and a true lover of nature. He did great service to botany by supplying plants and seeds to Linnæus, Dillenius, Collinson, and other European botanists. He was a collector, however, rather than an investigator, and his successes seem to have been due, in the main, to the patient promptings and advice of his friend Collinson in London. Garden, whom he visited at Charleston in 1765, after his appointment as King's Botanist wrote of him to Ellis:

I have been several times into the country, and places adjacent to town, with him, and have told him the classes, genera, and species of all the plants that occurred, which I knew. I did this in order to facilitate his enquiries, as I find he knows nothing of the generic characters of plants, and can neither class them nor describe them; but I see that, from great natural strength of mind and long practice, he has much acquaintance with the specific characters; though this knowledge is rude, inaccurate, indistinct, and confused, seldom determining well between species and varieties. He is, however, alert, active, industrious, and indefatigable in his pursuits.[85]

Fothergill says in his Memoir of Collinson "that the eminent naturalist, John Bartram, may almost be said to have been created by my friend's assistance."

The foregoing remarks concerning the elder Bartram are simply for the purpose of calling attention to his proper position among the American naturalists of his day. It is not that I esteem Bartram the less, but that I esteem Garden, Clayton, Mitchell, and Colden more. The name of Bartram brings up at once that of his friend and patron, Peter Collinson, just as that of Garden reminds us of John Ellis.

Collinson and Ellis were never in America, yet if any men deserve to be called the fathers of American natural history it is they. For a period of thirty years or more, that period during which Linnæus was bringing about those reforms which have associated his name forever with the history of the classificatory sciences, these enlightened the science-loving London merchants seem to have held the welfare of American science in their keeping and to have faithfully performed their trust. I know few books which are more delightful than Darlington's Memorial or Bartram and Smith's Correspondence of Linnæus, made up as they are largely of the letters which passed between Collinson and Ellis and their correspondents in America, and with Linnæus, to whom they were constantly transmitting American notes and specimens.[86]

Humphrey Marshall [b. 1722, d. 1801] was a farmer-botanist of the Bartram type, and the author of The American Grove, a treatise upon the forest trees and shrubs of the United States, the first botanical work which was entirely American. Darlington's Memorials of Bartram and Marshall is a worthy tribute to this useful man.

Moses Bartram, a nephew of John, was also a botanist, and William, his son [b. 1739, d. 1823], was a much more prominent figure in American science. His Travels through North and South Carolina, published in 1791, was, in the opinion of Coues, the starting point of the distinctively American school of ornithology.

Collinson was a correspondent of Benjamin Franklin, and is said not only to have procured and sent to him the first electrical machine which came to America, but to have made known to him in 1743 the results of the first experiments in electricity, the continuation of which gave to Franklin his European reputation as a man of science. Collinson was instrumental in introducing grape culture in Virginia, and in acclimating here many foreign ornamental shrubs.

Ellis was a more eminent man of science, and his name is associated with the beginnings of modern marine zoology.

Linnæus wrote to him in 1760: "Your discoveries may be said to vie with those of Columbus. He found out America, or a new India, in the west; you have laid open hitherto unknown Indies in the depths of the ocean." He was royal agent for West Florida, and had extraordinary facilities for obtaining specimens from the colonies.

His nephew, Henry Ellis, F. R. S. [b. 1720, d. 1805], was the author of A Voyage to Hudson's Bay in 1746 and 1747 for Discovering a North

West Passage, which contains some valuable notes upon zoology. He was in 1756 appointed governor of the colony of Georgia, and in 1758 published in the Philosophical Transactions an essay on the Heat of the Weather in Georgia. In 1760 he made a voyage for the discovery of a new passage to the Pacific, and later was governor of Nova Scotia, where we can but believe he continued his observations and his correspondence with the savans of Europe. "Finally," says Jones, "having attained a venerable age, and to the last intent upon the prosecution of some favorite physical researches, he fell in sleep, as did Pliny the Elder, within sight of Vesuvius, and upon the shores of the beautiful Bay of Naples." [87]

Jones, in his History of Georgia [I, p. 444], refers to the Rev. Stephen Hales—"equally renowned as a naturalist and a divine"—who lived for a time in Georgia during the last century. Can this have been the famous author of Vegetable Statics? I have been unable to find any allusion to a sojourn in America, in the published notices of the English Hales, and equally unable to discover a second Hales in the annals of science.

The central figure among eighteenth-century naturalists was of course Linnæus. His Systema Naturæ was an epoch-making work, and with the publication of its first edition at Leyden in 1735 the study of the biological sciences received an impress which was soon felt in America.

In 1738, while in Leyden, he assisted Gronovius in editing the notes sent by Clayton from Virginia, and it is evident that Linnæus was already, at the age of thirty, recognized by European botanists as an authority upon the plants of America. It was in this year that he visited Paris. He at once made his way to the Garden of Plants, and entered the lecture room of Bernard de Jussieu, who was describing some exotics to his pupils in Latin. There was one which the demonstrator had not yet determined, and which seemed to puzzle him. The Swede looked on in silence at first, but observing the hesitation of the learned professor, cried out: "Haec plantam faciem Americanam habet." Jussieu turned about quickly with the exclamation, "You are Linnæus."

It is interesting to notice how strongly the Linnæan reforms took root in American soil, and how soon. Collinson wrote to Bartram in 1737: "The Systema Naturæ is a curious performance for a young man, but his coining a new set of names for plants tends but to embarrass and perplex the study of botany. As to his system . . . botanists are not agreed about it. Very few like it. Be that as it will, he is certainly a very ingenious man, and a great naturalist."[88] Six years later he wrote to Linnæus himself:

Your system, I can tell you, obtains much in America. Mr. Clayton and Dr. Colden at Albany on Hudson's River in New York, are complete Professors; as is Dr. Mitchell at Urbana on Rappahannock River, in Virginia.[89]

This may not seem a very numerous following, but twelve years after this (1755) only seven English botanists were mentioned by Collinson in response to a request from Linnæus to know what botanical people in London were skilled in his plan.[90]

It is a fact not often referred to that during his period of poverty and struggles, Linnæus received, through the influence of his patron, Boerhaave, an appointment in the colony of Surinam. His prospects for a successful career in Europe had, however, brightened, and he decided not to come to America.

His interest in American natural history was always very great, and his descriptions of New World forms seem to have been drawn up with especial care. Garden, Colden, Bartram, Mitchell, Clayton, and Ellis were all, as we have seen, active in supplying him with materials, and his pupils, Kalm, Alstroem, Loefling, Kuhn, and Rolander (who collected for many years in Surinam) sent him many notes and specimens.

The progress of systematic zoology in the interval between Ray and Linnæus may perhaps best be illustrated by some brief statistical references. The former, in 1690, made an estimate of the number of animals and plants known at that time.

The numbers of beasts, including serpents, he placed at 150, adding that according to his belief not many that are of any considerable bigness in the known regions of the world have escaped the cognizance of the curious.

Linnæus in his twelfth edition (1766) described 210 species of beasts or mammals, and 124 of reptiles, so called. Of the mammals known to Linnæus, 78, or more than one-third were American, and 88 of the reptiles were attributed to this country.

"The number of birds," said Ray, "may be near 500. "Linnæus catalogued 790, of which about one-third were American.

Although at this time the Middle and Southern States were the most active in the prosecution of scientific researches, there were in New England at least two diligent students of nature. Paul Dudley, F. R. S. [b. 1675], chief justice of the colony of Massachusetts, was the author of several papers in the Philosophical Transactions. Among these were A description of the Moose Deer in America,[91] An Account of a Method lately found out in New England for Discovering where the Bees Hive in the Woods,[92] An Account of the Rattlesnake,[93] and An Essay upon the Natural History of Whales, with a particular Account of the Ambergris found in the Spermaceti Whale,[94] which is often quoted.

Others were An account of the Poyson Wood Tree in New England,[95] and Observations on some Plants in New England, with remarkable Instances of the Nature and Power of Vegetation.[96] He also appears to have sent to Collinson a treatise upon the evergreens of New England.[97]

The Rev. Jared Eliot [b. 1685, d. 1763], minister at Killingworth, in Connecticut, and one of the earliest graduates of Yale College, described by his contemporaries as "the first physician of his day," and as "the first botanist in New England," appears to have been a correspondent of Franklin and a scientific agriculturist.

In 1781 appeared Jefferson's Notes on Virginia. This was the first comprehensive treatise upon the topography, natural history, and natural resources of one of the United States, and was the precursor of the great library of scientific reports which have since been issued by the State and Federal Governments.

The book, although hastily prepared to meet a special need, and not put forth as a formal essay upon a scientific topic, was, if measured by its influence, the most important scientific work as yet published in America. The personal history and the public career of Thomas Jefferson are so familiar to all that it would be an idle task to repeat them here. Had he not been a master in statecraft he would have been a master of science. It is probable that no two men have done so much for science in America as Jefferson and Agassiz—not so much by their direct contributions to knowledge as by the immense weight which they gave to scientific interests by their advocacy.

Many pages of Jefferson's Notes on Virginia are devoted to the discussion of Buffon's statements: (1) That the animals common to both continents are smaller in the New World; (2) that those which are peculiar to the New are on a smaller scale; (3) that those which have been domesticated in both have degenerated in America, and (4) that, on the whole, America exhibits fewer species. He successfully overthrows the specious and superficial arguments of the eloquent French naturalist, who, it must be remembered, was at this time considered the highest authority living in such matters. Not content with this, when minister plenipotentiary to Europe a few years later he forced Buffon himself to admit his error.

The circumstance shall be related in the words of Daniel Webster, who was very fond of relating the anecdote:

It was a dispute in relation to the moose, and in one of the circles of the *beauxesprits* in Paris, Mr. Jefferson contended for some characteristics in the formation of the animal, which Buffon stoutly denied. Whereupon Mr. Jefferson wrote from Paris to General John Sullivan, then residing in Durham, New Hampshire, to procure and send him the whole frame of a moose. The General was no little astonished at a request he deemed so extraordinary, but, well acquainted with Mr. Jefferson, he knew he must have sufficient reason for it, so he made a hunting party of his neighbors and took the field. They captured a moose of unusual proportions, stripped it to the bone, and sent the skeleton to Mr. Jefferson at a cost of £50. On its arrival Mr. Jefferson invited Buffon and some other savants to a supper at his house and exhibited his dear-bought specimen. Buffon immediately acknowledged his error. "I should have consulted you, Monsieur," he said, "before publishing my book on Natural History, and then I should have been sure of my facts."

In still another matter in which he was at variance with Buffon he was manifestly in the right. In a letter to President Madison, of William and Mary College, he wrote:

Speaking one day with M. de Buffon on the present ardor of chemical inquiry, he affected to consider chemistry but as cookery and to place the toils of the laboratory on a footing with those of the kitchen. I think it, on the contrary, among the most useful of sciences and big with future discoveries for the utility and safety of the human race.

It was the scientific foresight of Jefferson, so manifest in such letters, which led him to advocate so vigorously the idea that science must be the corner stone of our Republic.

In 1789 he wrote from Paris to Doctor Willard, president of Harvard College:

To Doctor WILLARD:

What a field have we at our doors to signalize ourselves in. The botany of America is far from being exhausted, its mineralogy is untouched, and its natural history of zoology totally mistaken and misrepresented. . . . It is for such institutions as that over which you preside so worthily, sir, to do justice to our country, its productions, and its genius. It is the work to which the young men you are forming should lay their hands. We have spent the prime of our lives in procuring them the precious blessing of liberty. Let them spend theirs in showing that it is the great parent of science and of virtue, and that a nation will be great in both always in proportion as it is free.

THOMAS JEFFERSON.

To Jefferson's interest was due the organization of the first Government exploring expedition. As early as 1780 we find him anxious to promote an expedition to the upper portion of the Mississippi Valley, and offering to raise 1,000 guineas for the purpose from private sources, and while he was President he dispatched Lewis and Clarke upon their famous expedition into the Northwest—the precursor of all the similar enterprises carried on by the General Government, which have culminated in our magnificent Geological Survey.

Jefferson's personal influence in favor of science was of incalculable value. Transferred from the presidency of the principal American scientific society to the Presidency of the nation, he carried with him to the Executive Mansion the tastes and habits of a scientific investigator. Mr. Luther, in his recent essay upon Jefferson as a Naturalist,[98] has shown that during his residence in Paris he kept the four principal colleges—Harvard, Yale, William and Mary, and the College of Philadelphia—informed of all that happened in the scientific circles of Europe.

He wrote to one correspondent: "Nature intended me for the tranquil pursuits of science, by rendering them my supreme delight." To another he said: "Your first gives me information in the line of natural history, and the second promises political news. The first is my passion, the last my duty, and therefore both desirable."

When Jefferson went to Philadelphia to be inaugurated Vice-President he carried with him a collection of fossil bones which he had obtained in Greenbrier County, West Virginia, together with a paper, in which were formulated the results of his study upon them. This was published in the Transactions of the American Philosophical Society, and the species is still known as *Megalonyx jeffersoni*.

"The spectacle," remarks Luther, "of an American statesman coming to take part as a central figure in the greatest political ceremony of our country and bringing with him an original contribution to the scientific knowledge of the world, is certainly one we shall not soon see repeated."[99]

When Jefferson became President his specific tastes were the subject of much ridicule as well as of bitter opposition among the people in whose eyes, even in that day, science was considered synonymous with atheism. William Cullen Bryant, then a lad of thirteen, wrote a satirical poem, The Embargo, since suppressed, in which the popular feeling seems to have been voiced:

> Go, wretch, resign the presidential chair,
> Disclose thy secret measures, foul of fair.
> Go, search with curious eyes for horned frogs,
> 'Mid the wild wastes of Louisianian bogs;
> Or, where the Ohio rolls his turbid stream,
> Dig for huge bones, thy glory and thy theme.

A prominent personage in the history of this period was Peter Kalm, a pupil of Linnæus and professor in the University of Aobo, who was sent to America by Swedish Government, and traveled through Canada, New York, New Jersey, and Pennsylvania from 1748 to 1751. Although the ostensible object of his mission was to find a species of mulberry suitable for acclimatization in Sweden, with a view to the introduction of silk culture, it is very evident that he and his master were very willing to make of applied science a beast of burden, upon whose back they could heap up a heavy burden of investigation in pure science. Kalm's botanical collections were of great importance and are still preserved in the Linnæan Herbarium in London. His Travels into North America are full of interesting observations upon animals and men, as well as upon plants, and give us an insight into the life of the naturalists at that time resident in America. After his return to Sweden he published several papers relating to his discoveries in America.

Another traveler who deserves our attention, Johann David Schœpf [b. 1752, d. in Baireuth, 1800], the author of one of the earliest monographs of the Testudinata, was a surgeon of mercenary troops under the Marcgrave of Anspach, and was one of the hated Hessian auxiliaries during the Revolutionary war (1776–1783). While stationed at New York he wrote a paper upon the Fishes of New York, which was published in Berlin in 1787. This was the first special ichthyological paper ever written in America or concerning American species. Immediately after the treaty of peace in 1783, Schœpf made an extensive tour through the United States, proceeding from New York south to Florida and the Bahamas. He was accompanied in his more southern excursions by Professor Marter and Doctor Stupicz, who, with several assistants, had been sent to America from Vienna to make botanical exploration. Schœpf's Nord Amerikanische Reisen is full of interesting notes upon natural history, and describes nearly all the scientific men at that time resident in the United States. His Materia Medica Americana, published in 1787 at Erlangen, was a standard in its day. [100]

One of the most prominent names in American natural history is that of John Reinhold Forster [b. 1729, d. 1798], who was a leader in zoological studies in England during the last century. He was a native of Germany, and at the time of his death professor of botany at Halle. He spent many years in England, and was the naturalist of Cooke's second voyage around the world (1772–1775). In 1771 he published in London, in an appendix to his translation of Kalm's Travels, A Catalogue of the Animals of North America, compiled from the writings of Linnæus, Pennant, Brisson, Edwards, and Catesby, and in the same year a similar nominal catalogue of the plants of North America. His account of the birds sent from Hudson Bay, published in 1772, was a valuable contribution to American ornithology, "notable," says Coues, "as the first formal treatise exclusively devoted to a collection of North American birds sent abroad." Fifty-eight species were described, among which were several new to science. Other papers of equal value were published upon the quadrupeds and fishes of the region. Forster was one of the earliest students of the geographical distribution of animals, and his Enchiridion of Natural History was in its day a standard. His son, John George Forster, who was his companion in the voyage of circumnavigation, owes his fame to his literary rather than to his scientific labors. He published a paper on the Patella or Limpet Fish found at Bermuda.[101]

The annals of Russian explorations upon the west coast of North America have been so exhaustively recorded by Dall in his Alaska and its Resources that only passing mention need be made of the two German naturalists, Steller and Chamisso, whose names are identified with natural history work of Russian explorer.

Among other naturalists whose names are associated with America during this period may be mentioned Sonnini de Manoncour, an eminent French zoologist, who traveled in Surinam from 1771 to 1775 and made important contributions to its ornithology. Don Felix de Azara [b. 1746, d. after 1806], who carried on researches in Spanish America from 1781 to 1801; Don Antonio Parra, who published a useful treatise on the natural history of Cuba in Havana, in 1787; Don Joseph C. Mutis, a learned Spanish ecclesiastic and physician, professor of natural history in the University of Santa Fe de Bogota, in Grenada, who carried on a voluminous correspondence with Linnæus and his son from 1763 to 1778,[102] and Joseph Jussieu, botanist to the King of France, who went to the west coast of South America in 1734 as a member of the commission sent to the Royal Academy of Sciences to make observations to determine more accurately the shape and magnitude of the earth. "His curiosity," says Flourens, "held him captive for many years in these regions so rich and unexplored, where he often joined the labors of the engineer with those of the botanist. To him Europe owes several new plants, the heliotrope, the marvel of Peru, etc., with many curious and then unknown species." Here, also should be mentioned the eminent French ornithologist, Francois Levaillant [b. 1753, d. 1824], who was a native of America, and the two Mexican naturalists, also native born, Jose A. Alzate [b. in Ozumba 1729, d. in Mexico February 2, 1790], a learned botanist, and Francisco Xavier Clavigero.

Francisco Xavier Clavigero, the historian of Mexico, was one of the earliest of American archæologists. Born in Vera Cruz September 9, 1731, the son of a Spanish scholar, he was educated at the college of Puebla, entered the Society of Jesus, and was sent out as a missionary among the Indians, with whom he spent thirty-six years. He learned their language, collected their traditions, and examined all their historical records and monuments for the purpose of correcting the misrepresentations of early Spanish writers. When the Society of Jesus was suppressed by Spain, in 1767, Clavigero went to Italy, where he wrote his Storia Antica del Messico, printed in 1780–81.

Clavigero was a man who, in his spirit, was fully abreast of the science of his day, but whose methods of thought and argument were already antiquated.

His monastic training led him to write from the standpoint of a commentator rather than that of an original observer, and his observations upon the animals and plants of Mexico were subordinated in a very unfortunate manner to those of his predecessor, Hernandez. In the Dissertations, which make up the fourth volume of his history, he throws aside, in the ardor of his dispute with Buffon and his followers, the trammels of tradition, and places upon record many facts concerning American natural history which had never before been referred to. He here presented a list of the quadrupeds of America, the first ever printed for the entire continent, including 143 species; not systematically arranged, it is true, but perhaps as scientific in its construction as was possible at that time, even had its author been trained in the school of Linnæus.

Clavigero's dissertations are well worthy of the attention of naturalists even of the present day. His essay upon the manner in which the continent of America was peopled with living forms, shows a remarkable appreciation of the difficulties in the way of the solution of this still unsolved problem. The position taken by its author is not unlike that held by zoogeographers of to-day, in considering it necessary to bridge with land the water between Asia and Northwestern America, and Africa and South America.[103] In his first Dissertation of the Animals of Mexico he combats the prevailing European views as to the inferiority of the soil and climate of the New World and the degeneracy of its inhabitants, engaging in the same battle in which fought also Harriot, Acosta, and Jefferson.

Clavigero's contributions to archæology and ethnology are extensive and valuable, and we can but admit that at the time of the issue of his Storia Antica no work concerning America had been printed in English which was equally valuable.

Although in his formal discussion of the natural history of Mexico he follows closely the nomenclature and arrangement of Hernandez, there are many important original observations inserted. I will instance only the the notes on the mechanism of the poison gland and fang of the rattlesnake, the biographies of the possum, the coyote and the tapir, and the Tuza or pouched rat, the mocking bird, the chegoe, and the cochineal insect. Clavigero states that Father Inamma, a Jesuit missionary of California, has made many experiments upon snakes which serve to confirm those made by Mead upon vipers.

To the post-Revolutionary period belongs Doctor Manasseh Cutler, for fifty-one years minister of Ipswich Hamlet, Massachusetts [b. 1743, d. 1823], who in 1785 published An Account of some of the vegetable Productions, naturally growing in this Part of America, botanically arranged,[104] in which he described about 370 species. Cutler was a correspondent of Muhlenberg in Pennsylvania, Swartz and Payshull in Sweden, and Withering and Stokes in England. He left unpublished manuscripts of great value. He was one of the founders of the settlement in Ohio, and at one time a member of Congress. After Cutler, says Tuckerman, there appeared in the Northeastern States nothing of importance until the new school of New England botanists, a school characterized by the names of an Oakes, a Boott, and an Emerson, was founded in 1814, by the publication of Bigelow's Florula Bostoniensis.

Thomas Walter [b. in Hampshire, 1740] published in London, in 1787, his Flora Caroliniana, a scholarly work describing the plants of a region situate upon the Santee River.[105]

Doctor Hugh Williamson, of North Carolina [b. 1735, d. 1819], was a prominent member of the American Philosophical Society. He was concerned in some of the earliest astronomical and mathematical work in America; published papers upon comets and climatology, which were favorably received, and secured his election to many foreign societies, and in 1775 printed in the Philosophical Transactions his Experiments and Observations on the *Gymnotus Electricus*, or Electrical Eel.

Doctor Caspar Wistar [b. 1761, d. 1818] was one of the early professors of chemistry [1789] and anatomy [1793] in the College of Philadelphia. He was the discoverer of some important points in the structure of the ethmoid bone, a man of eminence as a teacher, and versed in all the sciences of his day.

Doctor James Woodhouse, of Philadelphia [b. 1770, d. 1809], made investigations in chemistry, mineralogy, and vegetable physiology which were considered of importance.

The story of the origin of American scientific societies has been so often told that it need not be repeated here. The only institutions of the kind which were in existence at the end of the period under consideration were the American Philosophical Society, an outgrowth primarily of the American Society for the Advancement of Natural Knowledge, founded in Philadelphia in 1743, and secondarily of Franklin's famous Junto, whose origin dates back to 1727, and the American Academy of Arts and Sciences, founded in 1780.

The relations of the colonial naturalists to the scientific societies of England have not so often been referred to, and it does not seem to be generally known that the early history of the Royal Society of London was intimately connected with the foundation of New England, and that the first proposition for the establishment of a scientific society in America was under consideration early in the seventeenth century. "The great Mr. Boyle," writes Eliot, "Bishop Wilkins, and several other learned men, had proposed to leave England and establish a society for promoting natural knowledge in the new colony, of which Mr. Winthrop, their intimate friend and associate, was appointed governor. Such men were too valuable to lose from Great Britain; and Charles II having taken them under his protection, the society was there established, and obtained the title of the Royal Society of London." [106]

For more than a hundred years the Royal Society was the chief resource of naturalists in North America. The three Winthrops, Mitchell, Clayton, Garden, Franklin, Byrd, Rittenhouse, and others were among its fellows, and the Philosophical Transactions contained many American papers.

As at an early date the Society of Arts in London began to offer prizes for various industrial successes in the colonies, for instance, for the production of potash and pearlash, for the culture of silk, and for the culture of hemp, the vine, safflower, olives, logwood, opium, scammony, burilla, aloes, sarsaparilla, cinnamon, myrtle wax, the production of saltpeter, cobalt, cochineal, the manufacture of wine, raisins, and olive oil, the collection of gum from the persimmon tree, and the acclimation of silk grass. A medal was given in 1861 to Doctor Jared Eliot, of Connecticut, for the extraction of iron from "black sand." [107] In 1757 we find their secretary endeavoring to establish branch societies in the colonial cities, especially in Charleston, Philadelphia, and New York, and Garden seems to have tried to carry out the enterprise in Charleston. After two years he wrote that the society organized had become "a mere society of drawing, painting, and sculpture."

In a subsequent letter he utters a pitiful plaint. He has often wondered, he says "that there should be a country abounding with almost every sort of plant, and almost every species of the animal kind, and yet that it should not have pleased God to raise up one botanist." [108]

The American Academy of Arts and Sciences was founded by the legislature of Massachusetts in 1780, and its first volume of memoirs appeared in 1785.

In 1788 an effort was made by the Chevalier Quesnay de Beaurepaire to found in Richmond, Virginia, the Academy of Arts and Sciences of the United States of America, upon the model of the French Academy. The plan was submitted to the Royal Academy of Sciences in Paris, and received its unqualified indorsement, signed, among others, by Lavoisier. A large subscription was made by the Virginians and a large building erected, but an academy of sciences needs members as well as a president, and the enterprise was soon abandoned. [109]

In 1799 was organized the Connecticut Academy of Arts and Sciences, which, after publishing one volume of Transactions, went into a state of inactivity from which it did not arouse itself until 1866.

This sketch would not be complete without some reference also to the history of scientific instruction in America during the last century.

The first regular lectures upon a special natural history topic appear to have been upon comparative anatomy. A course upon this topic was delivered at Newport, Rhode Island, in 1754, by Doctor William Hunter, a native of Scotland [b. about 1729], a kinsman of the famous English anatomists, William and John Hunter, and a pupil of Munro. His course upon comparative anatomy was given in connection with others upon human anatomy and the history of anatomy, the first medical lectures in America.[110]

The first instruction in botany was given in Philadelphia in 1768 by Kuhn, who began in May of that year a course of lectures upon that subject in connection with his professorship of materia medica and botany in the College of Philadelphia. Adam Kuhn [b. in Germantown, Pennsylvania, 1741, d. 1817] was educated in Europe, and had been a favorite pupil of Linnæus. He did not, however, continue his devotion to natural history, though he became an eminent physician. William Bartram, son of John Bartram, was elected to the same professorship in 1782. In 1788 Professor Waterhouse, of Harvard College, read lectures upon natural history to his medical classes, and is said to have subsequently claimed that these were the first public lectures upon natural history given in the United States. This was doubtless an error, for we find that in 1785 a course upon the philosophy of chemistry and natural history was delivered in Philadelphia. "People of every description, men and women, flock to these lectures," writes a contemporary. "They are held at the university three evenings in a week."[111]

The first professor of chemistry was Doctor Benjamin Rush, who lectured in the Philadelphia Medical School as early as 1769. Bishop Madison was professor of chemistry and philosophy at William and Mary College from 1774 to 1777; Aaron Dexter, of chemistry and materia medica at Harvard, 1783 to 1816; John Maclean, at Princeton, 1795–1812, being the first to occupy a separate chair of chemistry. Before the days of chemical professorships, the professor of mathematics seems to have been the chief exponent of science in our institutions of learning.

John Winthrop [b. 1714, d. 1779], for instance, who was Hollis professor of mathematics and natural philosophy at Harvard from 1738 to 1779, was a prominent Fellow of Royal Society, to whose Transactions he communicated many important papers, chiefly astronomical. We read, however, that Count Rumford imbibed from his lectures his love for physical and chemical research, and from this it may be inferred that he taught as much of chemistry as was known in his day. William Small, professor of mathematics in William and Mary from 1758 to 1762, was a man of similar tastes, though less eminent. He was the intimate friend of Eramus Darwin. President Jefferson was his pupil, attended his lectures on natural philosophy, and got from time to time his "first views of the expansion of science and of the system of things in which we are placed."

Doctor Samuel Latham Mitchill [b. 1764, d. 1831] was the first man to hold a professorship of natural history, lecturing upon that subject, together with chemistry, in Columbia College in 1792. Doctor Mitchill was eminent as a zoologist, mineralogist, and chemist, and not only published many valuable papers, but in 1798 established the first American scientific journal.

Harvard appears to have had the first separate professorship of natural history, which was filled by William Dandridge Peck, a zoologist and botanist of prominence in his day.

A professorship of botany was established in Columbia College, New York, as early as 1795, at which time Doctor David Hosack [b. in New York, 1769, d. 1835] was the incumbent. Doctor Hosack brought with him from Europe, in 1790, the first cabinet of minerals ever seen in the United States. In its arrangement he was assisted by one of his pupils, Archibald Bruce, who became, in 1806, professor of mineralogy, and who, soon after, in 1810, established the American Journal of Mineralogy.

Doctor Hosack was the founder of the first public botanic garden— this was in New York in 1801; another was founded in Charleston in 1804. This had disappeared forty years ago, and one at Cambridge, established in 1808, is the only one now in existence.

The first public museum was that founded in Philadelphia, in 1785, by Charles Willson Peale, the bones of a mammoth and a stuffed paddlefish forming its nucleus. This establishment had a useful career of nearly fifty years.

VII

We have now rehearsed the story of the earliest investigators of American natural history, including two centuries of English endeavor, and nearly three if we take into consideration the earlier explorations of the naturalists of continental Europe. We have seen how, in the course of many generations, the intellectual supremacy of the Western Continent went from the Spaniards and the French and the Dutch to the new people who were to be called Americans, and we have become acquainted with the men who were most thoroughly identified with the scientific endeavors of each successive period of activity.

The achievements of American science during the century which has elapsed since the time when Franklin, Jefferson, Rittenhouse, and Rumford were its chief exponents have been often the subject of presidential addresses like this, and the record is a proud one. During the last fifty years in England, and the last forty in America, discovery has followed discovery with such rapid succession that it is somewhat hard to realize that American science in the colonial period, or even that of Europe at the same time, had any features which are worthy of consideration.

The naturalists whose names I have mentioned were the intellectual ancestors of the naturalists of to-day. Upon the foundations which they laid the superstructure of modern natural history is supported. Without the encyclopedists and explorers there could have been no Ray, no Klien, no Linnæus. Without the systematists of the latter part of the eighteenth century the school of comparative anatomists would never have arisen. Had Cuvier and disciples never lived there would have been no place for the philosophic biologists of to-day.

The spirit of the early naturalists may be tested by passages in their writings which show how well aware they were of the imperfections of their work. Listen to what John Lawson, the Carolina naturalist, wrote in the year 1700:

The reptiles or smaller insects are too numerous to relate here, this country affording innumerable quantities thereof; as the flying stags with horns, beetles, butterflies, grasshoppers, locusts, and several hundreds of uncouth shapes, which in the summer season are discovered here in Carolina, the description of which requires a large volume, which is not my intent at present; besides, what the mountainous part of this land may hereafter open to our view, time and industry will discover, for we that have settled but a small share of this large province can not imagine, but there will be a great number of discoveries made by those that shall come hereafter into the back part of this land, and make inquiries therein, when, at least, we consider that the westward of Carolina is quite different in soil, air, weather, growth of vegetables, and several animals, too, which we at present are wholly strangers to, and seek for. As to a right knowledge thereof, I say, when another age is come, the ingenious then in being may stand upon the shoulders of those that went before them, adding their own experiments to what was delivered down to them by their predecessors, and then there will be something toward a complete natural history, which, in these days, would be no easy undertaking to any author that writes truly and compendiously as he ought to do.

Herbert Spencer, in his essay on The Genesis of Science, lays stress upon the fact that the most advanced sciences have attained to their present power by a slow process of improvement, extending through thousands of years, that science and the positive knowledge of the uncultured can not be separated in nature, and that the one is but a perfected and extended form of the other. "Is not science a growth?" says he. "Has not science its embryology? And must not the neglect of its embryology lead to a misunderstanding of the principles of its evolution and its existing organization?"

It seems to me unfortunate, therefore, that we should allow the value of the labors of our predecessors to be depreciated, or to refer to the naturalists of the last century as belonging to the unscientific or to the archaic period. It has been frequently said by naturalists that there was no science in America until after the beginning of the present century. This is, in one sense, true; in another very false. There were then, it is certain, many men equal in capacity, in culture, in enthusiasm, to naturalists of to-day, who were giving careful attention to the study of precisely the same phenomena of nature. The misfortune of men of science in the year of 1785 was that they had three generations fewer of scientific predecessors than have we. Can it be doubted that the scientists of some period long distant will look back upon the work of our own time as archaic and crude, and catalogue our books among the "curiosities of scientific literature?"

Is it not incumbent upon workers in science to keep green the memory of those whose traditions they inherited? That it is, I do most steadfastly believe, and with this purpose I have taken advantage of the tercentenary of American biology to read this review of the work of the men of old.

Monuments are not often erected to men of science. More enduring, however, than monuments are those living and self-perpetuating memorials, the plants and animals which bear the names of the masters who knew them and loved them. Well have the Agassizs remarked that "there is a world of meaning hidden under our zoological and botanical nomenclature, known only to those who are intimately acquainted with the annals of scientific life in its social as well as its professional aspect." [112]

I hope I am not at this day entirely alone in my appreciation of the extreme appropriateness of this time-honored custom, although I know that many of our too matter-of-fact naturalists are disposed to abandon it, and that it is losing much of its former significance. In fact, in these days of unstable nomenclature, such tributes are often very evanescent. It seems fortunate that the names of some of the most honored of the early naturalists are perpetuated in well-established generic and specific combinations.[113]

When I see the *Linnæa borealis*, I am always reminded of the sage of Upsala, as he is represented in the famous Amsterdam painting, clad in Lapland fur, and holding a spray of that graceful arctic plant. *Magnolia* and *Wistaria* call up the venerable professors of botany at Montpelier and Philadelphia. *Tradescantia virginica* reminds me of John Tradescant and the Ashmolean Museum, whose beginnings were gathered by him in Virginia. The cape jessamine (*Gardenia*), the spring beauty (*Claytonia*), the partridge berry (*Mitchella*), the iron weed (*Vernonia*), the *Quercus bartramii* (= *Q. heterophylla*), the *Scarus catesbyi*, *Thalictrum* and *Asclepias cornuti*, *Macrurus fabricii*, *Didelphys* and *Canis azaræ*, *Chauliodus sloanei*, *Alutera schœfii*, *Stema forsteri*, *Stolephorus mitchilli*, *Malacanthus plumieri*, *Salix cutleri*, and *Pinus banksiana*, the *Kalmia*, the *Jeffersonia*, the *Hernandia*, the *Comptonia*, the *Sarracenia*, the *Gaultheria*, the *Kuhnia*, the *Ellisia*, the *Coldenia*, the *Robinia*, the *Banisteria*, the *Plumieria*, the *Collinsonia*, the *Bartramia*, all bear the names of men associated with the beginnings of natural history in America.

Yet, pleasant as it is to recall in such manner the achievements of the fathers of natural history, let us not do them the injustice to suppose that posthumous fame was the object for which they worked. Like Sir Thomas Browne, they believed that "the world was made to be inhabited by beasts, but to be studied by man." Let us emulate their works and let us share with them the admonitions of the Religio Medici.

"The wisdom of God," says Browne, "receives small honor from those vulgar heads that rudely stray about, and with a gross rusticity admire His works; those highly magnify Him whose judicious inquiry into His acts, and deliberate research into His creatures, return the duty of a devout and learned admiration. Therefore," he continues—

> Search while thou wilt and let thy reason go
> To ransom truth, even to the abysse below,
> Rally the scattered causes, and that line
> Which nature twists be able to untwine.
> It is thy Maker's will, for unto none
> But unto reason can He e'er be known.

THE BEGINNINGS OF
AMERICAN SCIENCE
THE THIRD CENTURY[1]

By George Browne Goode

President of the Biological Society of Washington

VIII

In the address which it was my privilege one year ago to read in the presence of this society I attempted to trace the progress of scientific activity in America from the time of the first settlement by the English in 1585 to the end of the Revolution, a period of nearly two hundred years.

Resuming the subject, I shall now take up the consideration of the third century, from 1782 to the present time. For convenience of discussion the time is divided, approximately, into decades, while the decades naturally fall into groups of three. From 1780 to 1810, from 1810 to 1840, from 1840 to 1870, and from 1870 to the close of the century are periods in the history of American thought, each of which seems to be marked by characteristics of its own. These must have names, and it may not be inappropriate to call the first the period of Jefferson, the second that of Silliman, and the third that of Agassiz.

The first was, of course, an extension of the period of Linnæus, the second and third were during the mental supremacy of Cuvier and Von Baer and their schools, and the fourth or present, beginning in 1870, belongs to that of Darwin, the extension of whose influence to America was delayed by the tumults of the civil convulsion which began in 1861 and ended in 1865.

The beginnings of American science do not belong entirely to the past. Our science is still in its youth, and in the discussion of its history I shall not hesitate to refer to institutions and to tendencies which are of very recent origin.

It is somewhat unfortunate that the account book of national progress was so thoroughly balanced in the centennial year. It is true that the movement which resulted in the birth of our Republic first took tangible form in 1776, but the infant nation was not born until 1783, when the treaty of Paris was signed, and lay in swaddling clothes until 1789, when the Constitution was adopted by the thirteen States.

In those days our forefathers had quite enough to do in adapting their lives to the changed conditions of existence. The masses were struggling for securer positions near home or were pushing out beyond the frontiers to find dwelling places for themselves and their descendants. The men of education were involved in political discussions as fierce, uncandid, and unphilosophical in spirit as those which preceded the French Revolution of the same period.

The master minds were absorbed in political and administrative problems and had little time for the peaceful pursuits of science, and many of the men who were prominent in science—Franklin, Jefferson, Rush, Mitchill, Seybert, Williamson, Morgan, Clinton, Rittenhouse, Patterson, Williams, Cutler, Maclure, and others—were elected to Congress or were called to other positions of official responsibility.

IX

The literary and scientific activities of the infant nation were for many years chiefly concentrated at Philadelphia, until 1800 the Federal capital and largest of American cities. Here, after the return of Franklin from France in 1785, the meetings of the American Philosophical Society were resumed. Franklin continued to be its president until his death in 1790, at the same time holding the presidency of the Commonwealth of Pennsylvania and a seat in the Constitutional Convention. The prestige of its leader doubtless gave to the society greater prominence than its scientific objects alone would have secured.

In the reminiscences of Doctor Manasseh Cutler there is to be found an admirable picture of Franklin in 1787. As we read it we are taken back into the very presence of the philosopher and statesman, and can form a very clear appreciation of the scientific atmosphere which surrounded the scientific leaders of the post-Revolutionary period.

Doctor Cutler wrote:

Doctor Franklin lives in Market Street, between Second and Third Streets, but his house stands up a court-yard at some distance from the street. We found him in his garden, sitting upon a grass plat under a large mulberry tree, with several other gentlemen and two or three ladies. When Mr. Gerry introduced me he rose from his chair, took me by the hand, expressed his joy to see me, welcomed me to the city, and begged me to seat myself close by him. His voice was low, his countenance open, frank, and pleasing. I delivered him my letters. After he had read them he took me again by the hand and, with the usual compliments, introduced me to the other gentlemen, who were, most of them, members of the Convention. Here we entered into a free conversation, and spent the time most agreeably until it was dark. The tea table was spread under the tree, and Mrs. Bache, who is the only daughter of the Doctor and lives with him, served it out to the company.

The Doctor showed me a curiosity he had just received, and with which he was much pleased. It was a snake with two heads, preserved in a large vial. It was about ten inches long, well proportioned, the heads perfect, and united to the body about one-fourth of an inch below the extremities of the jaws. He showed me a drawing of one entirely similar, found near Lake Champlain. He spoke of the situation of this snake, if it was traveling among bushes, and one head should choose to go on one side of the stem of a bush and the other head should prefer the other side, and neither of the heads would consent to come back or give way to the other. He was then going to mention a humorous matter that had that day taken place in the Convention, in consequence of his comparing the snake to America, for he seemed to forget that everything in the Convention was to be kept a profound secret; but this was suggested to him, and I was deprived of the story.

After it was dark we went into the house, and he invited me into his library, which is likewise his study. It is a very large chamber and high-studded. The walls were covered with shelves filled with books; besides, there were four large alcoves, extending two-thirds of the length of the chamber, filled in the same manner. I presume this is the largest and by far the best private library in America. He showed us a glass machine for exhibiting the circulation of the blood in the arteries and veins of the human body. The circulation is exhibited by the passing of a red fluid from a reservoir into numerous capillary tubes of glass, ramified in every direction, and then returning in similar tubes to the reservoir, which was done with great velocity, and without any power to act visibly upon the fluid, and had the appearance of perpetual motion. Another great curiosity was a rolling press for taking copies of letters or any other writing. A sheet of paper is completely copied in less than two minutes, the copy as fair as the original, and without effacing it in the smallest degree. It is an invention of his own, and extremely useful in many situations in life. He also showed us his long artificial arm and hand for taking down and putting up books on high shelves, out of reach, and his great armchair with rockers, and a large fan placed over it, with which he fans himself, while he sits reading, with only a slight motion of his foot, and many other curiosities and inventions, all his own, but of lesser note. Over his mantel-tree he has a prodigious number of medals, busts, and casts in wax or plaster of paris, which are the effigies of the most noted characters in Europe. But what the Doctor wished especially to show me was a huge volume on botany, which indeed afforded me the greatest pleasure of any one thing in his library. It was a single volume, but so large that it was with great difficulty that the Doctor was able to raise it from a low shelf and lift it to the table; but, with the senile ambition common to old people, he insisted on doing it himself, and would permit no one to assist him, merely to show how much strength he had remaining. It contained the whole of Linnæus Systema Vegetabilium, with large cuts of every plant colored from nature. It was a feast to me, and the Doctor seemed to enjoy it as well as myself. We spent a couple of hours examining this volume, while the other gentlemen amused themselves with other matters. The Doctor is not a botanist, but lamented that he did not in early life attend to this science. He delights in natural history, and expressed an earnest wish that I would pursue the plan I had begun, and hoped this science, so much neglected in America, would be pursued with as much ardor here as it is now in every part of Europe. I wanted, for three months at least, to have devoted myself entirely to this one volume, but, fearing I should be tedious to the Doctor, I shut the book, though he urged me to examine it longer. He seemed extremely fond, through the course of the visit, of dwelling on philosophical subjects, and particularly that of natural history, while the other gentlemen were swallowed up in politics. This was a favorable circumstance to me, for almost the whole of his conversation was addressed to me, and I was highly delighted with the extensive knowledge he appeared to have of every subject, the brightness of his memory, the clearness and vivacity of all his mental faculties. Notwithstanding his age (eighty-four), his manners are perfectly easy, and everything about him seems to diffuse an unrestrained freedom and happiness. He has an incessant vein of humor, accompanied with an uncommon vivacity, which seems as natural and involuntary as his breathing.

To Franklin, as president of the Philosophical Society, succeeded David Rittenhouse [b. 1732, d. 1796], a man of world-wide reputation, known in his day as *the* American philosopher.[2]

He was an astronomer of repute, and his observatory, built at Norriton in preparation of the transit of Venus in 1769, seems to have been the first in America. His orrery, constructed upon an original plan, was one of the wonders of the land. His most important contribution to astronomy was the introduction of the use of spider lines in the focus of transit instruments.[3]

He was an amateur botanist, and in 1770 made interesting physiological experiments upon the electrical eel.[4]

He was a Fellow of the Royal Society of London, and the first director of the United States Mint.

Next in prominence to Franklin and Rittenhouse were doubtless the medical professors, Benjamin Rush, William Shippen, John Morgan, Adam Kuhn, Samuel Powell Griffiths, and Caspar Wistar, all men of scientific tastes, but too busy in public affairs and in medical instruction to engage deeply in research, for Philadelphia, in those days as at present, insisted that all her naturalists should be medical professors, and the active investigators, outside of medical science, were not numerous. Rush, however, was one of the earliest American writers upon ethnology, and a pathologist of the highest rank. He is generally referred to as the earliest professor of chemistry, having been appointed to the chair of chemistry in the College of Philadelphia in 1769. It seems certain, however, that Doctor John Morgan lectured on chemistry as early as 1765.[5]

Doctor Shippen [b. 1735, d. 1808], the founder of the first medical school [1765] and its professor of anatomy for forty-three years, was still in his prime, and so was Doctor Morgan [b. 1735, d. 1789], a Fellow of the Royal Society, a co-founder of the medical school, and a frequent contributor to the Philosophical Transactions. Morgan was an eminent pathologist, and is said to have been the one to originate the theory of the formation of pus by the secretory action of the vessels of the part.[6] He appears to have been the first who attempted to form a museum of anatomy, having learned the methods of preparation from the Hunters and from Süe in Paris. The beginning was still earlier known, for a collection of anatomical models in wax, obtained by Doctor Abraham Chovet in Paris, was in use by Philadelphia medical students before the revolution.[7]

Another of the physicians of colonial days who lived until after the revolution was Doctor Thomas Cadwallader [b. 1707, d. 1779], whose dissections are said to have been among the earliest made in America, and whose Essay on the West India Dry Gripes, 1775, was one of the earliest medical treatises in America.

Doctor Caspar Wistar [b. 1761, d. 1818] was also a leader, and was at various times professor of chemistry and anatomy. His contributions to natural history were descriptions of bones of *Megalonyx* and other mammals, a study of the human ethmoid, and experiments on evaporation. He was long vice-president of the Philosophical Society, and in 1815 succeeded Jefferson in its presidency. The Wistar Anatomical Museum of the university and the beautiful climbing shrub *Wistaria* are among the memorials to his name.[8]

Still another memorial of the venerable naturalist may perhaps be worthy of mention as an illustration of the social condition of science in Philadelphia in early days. A traveler visiting the city in 1829 thus described this institution, which was continued until the late war and then discontinued, but has been resumed within the last year:

Doctor Wistar in his lifetime had a party of his literary and scientific friends at his house, one evening in every week—and to this party, strangers visiting the city, were also invited. When he died, the same party was continued, and the members of the Wistar party, in their tour, each have a meeting of the club at his house, on some Saturday night in the year. This club consists of the men most distinguished for learning, science, art, literature, and wealth in the city.

It opens at early candle-light, in the evening, where, not only the members themselves appear, but they bring with them all the strangers of distinction then in the city.[9]

The Wistar parties were continued up to the beginning of the civil war, in 1861, and have been resumed since 1887. A history of these gatherings would cover a period of three-quarters of a century at the least, and could be made a most valuable and entertaining contribution to scientific literature.

Packard, in his History of Zoology,[10] states that zoology, the world over, has sprung from the study of human anatomy, and that American zoology took its rise and was fostered chiefly in Philadelphia by the professors in the medical schools.

It was fully demonstrated, I think, in my former address, that there were good zoologists in America long before there were medical schools, and that Philadelphia was not the cradle of American natural history, although during its period of political preeminence, immediately after the Revolution, scientific activities of all kinds centered in that city. As for the medical schools, it is at least probable that they have spoiled more naturalists than they have fostered.

Doctor Adam Kuhn [b. 1741, d. 1817] was the professor of botany in 1768[11]—the first in America—and was labeled by his contemporaries the favorite pupil of Linnæus. Professor Gray, in a recent letter to the writer, refers to this saying as a myth; and it surely seems strange that a discipline beloved by the great Swede could have done so little for botany. Barton, in a letter, in 1792, to Thunberg, who then occupied the seat of Linnæus in the University of Upsala, said:

The electricity of your immortal Linné has hardly been felt in this *Ultima Thule* of science. Had a number of the pupils of that great man spread themselves along, and settled in the countries of North America, the riches of this world of natural treasures would have been better known. But alas! the one only pupil of your predecessor that had made choice of America as the place of his residence has added *nothing* to the stock of natural knowledge.[12]

The Rev. Nicholas Collin, rector of the Swedish churches in Pennsylvania, was a fellow-countryman and acquaintance of Linnæus[13] and an accomplished botanist, having been one of the editors of Muhlenberg's work upon the grasses, and an early writer on American linguistics. He read before the Philosophical Society, in 1789, An essay on those inquiries in natural philosophy which at present are most beneficial to the United States of North America, which was the first attempt to lay out a systematic plan for the direction of scientific research in America. One of the most interesting suggestions he made was that the Mammoth was still in existence.

The vast Mahmot [said he] is perhaps yet stalking through the western wilderness; but if he is no more, let us carefully gather his remains, and even try to find a whole skeleton of this giant, to whom the elephant was but a calf.[14]

General Jonathan Williams, U. S. A. [b. 1750, d. 1815], was first superintendent of the Military Academy at West Point and father of the Corps of Engineers. He was a nephew of Franklin and his secretary of legation in France, and, after his return to Philadelphia, was for many years a judge of the court of common pleas, his military career not beginning till 1801. This versatile man was a leading member of the Philosophical Society and one of its vice-presidents. His paper On the use of the thermometer in navigation was one of the first American contributions to scientific seamanship.

The Rev. Doctor John Ewing [b. 1732, d. 1802], also a vice-president, was provost of the university. He had been one of the observers of the transit in 1769, of which he published an account in the Transactions of the Philosophical Society. He early printed a volume of lectures on natural philosophy, and was the strongest champion of John Godfrey, the Philadelphian, in his claim to the invention of the reflecting quadrant.[15]

Doctor James Woodhouse [b. 1770, d. 1809] was author and editor of several chemical text-books and professor of chemistry in the university, a position which he took after it had been refused by Priestley. He made experiments and observations on the vegetation of plants and investigated the chemical and medical properties of the persimmon tree. He it was who first demonstrated the superiority of anthracite to bituminous coal by reason of its intensity and regularity of heating power.[16]

The Rev. Ebenezer Kinnersley [b. in Glouceser, England, November 30, 1711; d. in Philadelphia, July 4, 1778] survived the Revolution, though, in his latter years, not a contributor to science. The associate of Franklin in the Philadelphia experiments in electricity, his discoveries were famous in Europe as well as in America.[17] It is claimed that he originated the theory of the positive and negative in electricity; that he first demonstrated the passage of electricity through water; and that he first discovered that heat could be produced by electricity; besides inventing numerous mechanical devices of scientific interest. From 1753 to 1772 he was connected with the University of Pennsylvania, where there may still be seen a window dedicated to his memory.

Having already referred to the history of scientific instruction in America,[18] and shown that Hunter lectured on comparative anatomy in Newport in 1754; Kuhn on botany, in Philadelphia, in 1768; Waterhouse on natural history and botany, at Cambridge, in 1788; and some unidentified scholars upon chemistry and natural history, in Philadelphia, in 1785, it would seem unjust not to speak of Kinnersley's career as a lecturer. He seems to have been the first to deliver public scientific lectures in America, occupying the platform in Philadelphia, Newport, New York, and Boston, from 1751 to the beginning of the Revolution. The following advertisement was printed in the Pennsylvania Gazette for April 11, 1751:

NOTICE is hereby given to the *Curious*, That on *Wednesday* next, Mr. *Kinnersley* proposes to begin a Course of Experiments on the newly-discovered ELECTRI- CAL FIRE, containing not only the most curious of those that have been made and published in *Europe*, but a considerable Number of new Ones lately made in this City; to be accompanied with methodical LECTURES on the Nature and Properties of that wonderful Element.

Francis Hopkinson [b. 1737, d. 1791], signer of the Declaration of Independence, was treasurer of the Philosophical Society, and among other papers communicated by him was one in 1783, calling attention to the peculiar worm parasitic in the eye of a horse. The horse with a snake in its eye was on public exhibition in Philadelphia in 1782, and was the object of much attention, for the nature and habits of this peculiar *Filaria* were not so well understood then as now.

The father of Francis, Thomas Hopkinson [b. in London, 1709; d. in Philadelphia, 1751], who was overlooked in my previous address, de- serves at least a passing mention. Coming to Philadelphia in 1731, he became lawyer, prothonotary, judge of the admiralty, and member of the provincial council. As an incorporator of the Philadelphia Library Com- pany, and original trustee of the College of Philadelphia, and president of the first American Philosophical Society in 1743, his public spirit is wor- thy of our admiration. He was associated with Kinnersley and Franklin in the Philadelphia experiments, and Franklin said of him:

The power of points to throw off the electrical fire was first communicated to me by my ingenious friend, Mr. Thomas Hopkinson.[19]

The name of Philip Syng is also mentioned in connection with the Philadelphia experiments, and it would be well if some memorials of his work could be placed upon record.

William Bartram [b. 1739, d. 1824] was living in the famous botanical garden at Kingsessing, which his father, the old King's botanist, had bequeathed him in 1777. He was for some years professor of botany in the Philadelphia College, and in 1791 printed his charming volume descriptive of his travels in Florida, the Carolinas, and Georgia. The latter years of his life appear to have been devoted to quiet observation. William Bartram has been, perhaps, as much underrated as John Bartram has been unduly exalted. He was one of the best observers America has ever produced, and his book, which rapidly passed through several editions in English and French, is a classic, and should stand beside White's Selborne in every naturalist's library. Bartram was doubtless discouraged, early in his career, by the failure of his patrons in London to make any scientific use of the immense botanical collections made by him in the South before the Revolution, which many years later was lying unutilized in the Banksian herbarium. Coues has called attention very emphatically to the merits of his bird work, which he pronounces the starting point of a distinctly American school of ornithology. Two of the most eminent of our early zoologists, Wilson and Say, were his pupils; the latter, his kinsmen, and the former his neighbor, were constantly with him at Kingsessing and drew much of their inspiration from his conversation. Many birds which Wilson first fully described and figured were really named and figured by Bartram in his Travels, and several of his designations were simply adopted by Wilson.[20]

Bartram's Observations on the Creek and Cherokee Indians[21] was an admirable contribution to ethnography, and his general observations were of the highest value.

In the introduction to his Travels, and interspersed through this volume, are reflections which show him to have been the possessor of a very philosophic and original mind.

His Anecdotes of an American Crow and his Memoirs of John Bartram[22] were worthy products of his pen, while his illustrations to Barton's Elements of Botany show how facile and truthful was his pencil.

His love for botany was such, we are told, that he wrote a description of a plant only a few minutes before his death, a statement which will be readily believed by all who know the nature of his enthusiasm. Thus, for instance, he wrote of the Venus's Flytrap:

Admirable are the properties of the extraordinary Dionea muscipula! See the incarnate lobes expanding, how gay and sportive they appear! ready on the spring to intrap incautious, deluded insects! What artifice! There! behold one of the leaves just closed upon a struggling fly; another has gotten a worm; its hold is sure; its prey can never escape—carnivorous vegetable! Can we, after viewing this object, hesitate a moment to confess that vegetable beings are endowed with some sensible faculties or attributes, similar to those that dignify animal nature; they are organical, living, and self-moving bodies, for we see here, in this plant, motion and volition.[23]

Moses Bartram, a cousin of William, and also a botanist, was also living near Philadelphia, and in 1879 published Observations on the Native Silk Worms of North America, and Humphrey Marshall [1722–1801], the farmer-botanist, had a botanical garden of his own, and in 1785 published The American Grove—Arbustrium Americanum—a treatise on the forest trees and shrubs of the United States, which was the first strictly American botanical book, and which was republished in France a few years later, in 1789.

Gotthilf Muhlenberg [b. 1753, d. 1815], a Lutheran clergyman, living at Lancaster, was an eminent botanist, educated in Germany, though a native of Pennsylvania. His Flora of Lancaster was a pioneer work. In 1813 he published a full catalogue of the plants of North America, in which about 2,800 species were mentioned. He supplied Hedwig with many of the rare American mosses, which were published either in the Stirpes Cryptogamicæ of that author or in the Species Muscorum. To Sir J. E. Smith and Mr. Dawson Turner he likewise sent many plants. He made extensive preparations, writing a general flora of North America, but death interfered with his project. The American Philosophical Society preserves his herbarium, and the moss *Funeria muhlenbergii*, the violet *Viola muhlenbergii*, and the grass *Muhlenbergia* are among the memorials to his name.[24]

To Pennsylvania, but not to Philadelphia, came in 1794 Joseph Priestley [1733–1804], the philosopher, theologian, and chemist. Although his name is more famous in the history of chemistry than that of any living contemporary, American or European, his work was nearly finished before he left England. He never entered into the scientific life of the country which he sought as an exile, and of which he never became a citizen, and he is not properly to be considered an element in the history of American science.

His coming, however, was an event of considerable political importance, and William Cobbett's Observations on the Emigration of Doctor Joseph Priestley, by Peter Porcupine, was followed by several other pamphlets equally vigorous in expression. McMaster is evidently unjust to some of the public men who welcomed Priestley to America, though no one will deny that there were unprincipled demagogues in America in the year of grace 1794. Jefferson was undoubtedly sincere when he wrote to him the words quoted elsewhere in this address.

Another eminent exile welcomed by Jefferson, and the writer, at the President's request, of a work on national education in the United States, was M. Pierre Samuel Dupont de Nemours [b. in Paris, 1739; d. 1817]. He was a member of the Institute of France, a statesman, diplomatist, and political economist, and author of many important works. He lived in the United States at various times from 1799 till 1817, when he died near Wilmington, Delaware. Like Priestley, he was a member of the American Philosophical Society, and affiliated with its leading members.

The gunpowder works near Wilmington, Delaware, founded by his son in 1798, are still of great importance, and the statue of one of his grandsons, an Admiral in the United States Navy, adorns one of the principal squares in the national capital.

Among other notable names on the roll of the society in the last century were those of General Anthony Wayne and Thomas Payne. His Excellency General Washington was also an active member, and seems to have taken sufficient interest in the society to nominate for foreign membership the Earl of Buchan, president of the Society of Scottish Antiquarians, and Doctor James Anderson, of Scotland.

The following note written by Washington is published in the Memoirs of Rittenhouse:

The President presents his compliments to Mr. Rittenhouse, and thanks him for the attention he has given to the case of Mr. Anderson and the Earl of Buchan.

SUNDAY AFTERNOON, *20th April, 1794.*

Of all the Philadelphia naturalists of those early days the one who had the most salutary influence upon the progress of science was perhaps Benjamin Smith Barton [b. 1766, d. 1815]. Barton was the nephew of Rittenhouse and the son of Rev. Thomas Barton, a learned Episcopal clergyman of Lancaster, who was one of the earliest members of the Philosophical Society, and a man accomplished in science.

He studied at Edinburgh and Göttingen, and at the age of nineteen, in 1785, he was the assistant of Rittenhouse and Ellicott in the work of establishing the western boundary of Pennsylvania, and soon after was sent to Europe, whence, having pursued an extended course of scientific and medical study, he returned in 1789, and was elected professor of natural history and botany in the University of Pennsylvania. He was a leader in the Philosophical Society, and the founder of the Linnæan Society of Philadelphia, before which in 1807 he delivered his famous Discourse on some of the Principal Desiderata in Natural History, which did much to excite an intelligent popular interest in the subject. His essays upon natural history topics were the first of the kind to appear in this country. He belonged to the school of Gilbert White and Benjamin Stillingfleet, and was the first in America of a most useful and interesting group of writers, among whom may be mentioned John D. Godman, Samuel Lockwood, C. C. Abbott, Nicholas Pike, John Burroughs, Wilson Flagg, Ernest Ingersoll, the Rev. Doctor McCook, Hamilton Gibson, Maurice Thompson, and W. T. Hornaday, as well as Matthew Jones, Campbell Hardy, Charles Waterton, P. H. Gosse, and Grant Allen, to whom America and England both have claims.

Barton published certain descriptive papers, as well as manuals of botany and materia medica, but in later life had become so absorbed in medical affairs that he appears to have taken no interest in the struggles of the infant Academy of Natural Sciences, which was founded three years before his death, but of which he never became a member.

His nephew and successor in the presidency of the Linnæan Society and the University professorship, William P. C. Barton [b. 1786, d. 1856], was a man of similar tendencies, who in early life published papers on the flora of Philadelphia [Floræ Philadelphiæ Prodromus, 1815], but later devoted himself chiefly to professional affairs, writing copiously upon materia medica and medical botany.

The admirers of Benjamin Smith Barton have called him the father of American natural history, but the propriety of this designation, is questioned, since it is equally applicable to Mitchill or Jefferson, and perhaps still more so to Peter Collinson, of London. The praises of Barton have been so well and so often sung that there can be no injustice in passing him briefly by.[25]

The most remarkable naturalist of those days was Rafinesque [b. 1784, d. 1872], a Sicilian by birth, who came to Philadelphia in 1802.

Nearly fifty years ago this man died, friendless and impoverished in Philadelphia. His last words were these: Time renders justice to all at last. Perhaps the day has not yet come when full justice can be done to the memory of Constantine Rafinesque, but his name seems yearly to grow more prominent in the history of American zoology. He was in many respects the most gifted man who ever stood in our ranks. When in his prime he far surpassed his American contemporaries in versatility and comprehensiveness of grasp. He lived a century too soon. His spirit was that of the present period. In the latter years of his life, soured by disappointments, he seemed to become unsettled in mind, but as I read the story of his life his eccentricities seem to me the outcome of a boundless enthusiasm for the study of nature. The picturesque events of his life have been so well described by Jordan,[26] Chase,[27] and Audubon[28] that they need not be referred to here. The most satisfactory gauge of his abilities is perhaps his masterly Survey of the Progress and Actual State of Natural Sciences in the United States of America, printed in 1817.[29] His own sorrowful estimate of the outcome of his mournful career is very touching:

I have often been discouraged, but have never despaired long. I have lived to serve mankind, but have often met with ungrateful returns. I have tried to enlarge the limits of knowledge, but have ofted met with jealous rivals instead of friends. With a greater fortune I might have imitated Humboldt or Linnæus.

Doctor Robert Hare [b. 1781, d. 1858] began his long career of usefulness in 1801, at the age of twenty, by the invention of the oxyhydrogen blowpipe. This was exhibited at a meeting of the Chemical Society of Philadelphia in 1801.[30]

This apparatus was perhaps the most remarkable of his original contributions to science, which he continued without interruption for more than fifty years. It belongs to the end of the post-Revolutionary period, and is therefore noticed, although it is not the purpose of this essay to consider in detail the work of the specialists of the present century.

Doctor Hugh Williamson [b. December 5, 1753; d. in New York May 22, 1719] was a prominent but not particularly useful promoter of science, a writer rather than a thinker. His work has already been referred to. The names of Maclure, who came to Philadelphia about 1797, the Rev. John Heckewelder, and Albert Gallatin [b. 1761, d. in 1849], a native of Switzerland, a statesman and financier, subsequently identified with the scientific circles of New York, complete the list of the Philadelphia savants of the last century.

There is not in all American literature a passage which illustrates the peculiar tendencies in the thought of this period so thoroughly as Jefferson's defense of the country against the charges of Buffon and Raynal, which he published in 1783, which is particularly entertaining because of its almost pettish depreciation of our motherland.

On doit etre etonné [says Raynal] que l'Amerique n'ait pas encore produit un bon poëte, un habile mathematicien, un homme de génie dans un seul art, ou une seule science.

When we shall have existed as a people as long as the Greeks did before they produced a Homer, the Romans a Virgil, the French a Racine and Voltaire, the English a Shakespeare and Milton, should this reproach be still true, we will inquire from what unfriendly causes it has proceeded, that the other countries of Europe and quarters of the earth shall not have inscribed any name in the rôle of poets.

In war we have produced a Washington, whose memory will in future ages assume its just station among the most celebrated worthies of the world, when that wretched philosophy shall be forgotten which would have arranged him among the degeneracies of nature.

In physics we have produced a Franklin, than whom no one of the present age has made more important discoveries, nor has enriched philosophy with more, or more ingenious solutions of the phenomena of nature.

We have supposed Mr. Rittenhouse second to no astronomer living, that in genius he must be the first, because he is self-taught. He has not indeed made a world; but he has by imitation approached nearer its Maker than any man who has lived from the creation to this day. There are various ways of keeping the truth out of sight. Mr. Rittenhouse's model of the planetary system has the plagiary appellation of an orrery; and the quadrant invented by Godfrey, an American also, and with the aid of which the European nations traverse the globe, is called Hadley's quadrant.

We calculate thus: The United States contain three millions of inhabitants; France twenty millions; and the British Islands ten. We produce a Washington, a Franklin, a Rittenhouse. France then should have half a dozen in each of these lines, and Great Britain half that number, equally eminent. It may be true that France has; we are but just becoming acquainted with her, and our acquaintance so far gives us high ideas of the genius of her inhabitants.

The present war having so long cut off all communication with Great Britain, we are not able to make a fair estimate of the state of science in that country. The spirit in which she wages war is the only sample before our eyes, and that does not seem the legitimate offspring either of science or of civilization. The sun of her glory is fast descending to the horizon. Her philosophy has crossed the channel, her freedom the Atlantic, and herself seems passing to that awful dissolution whose issue is not given human foresight to scan.[31]

This was one phase of public sentiment. Another, no less instructive, is that shown forth in the publications of Jefferson's fierce political opponents in 1790, paraphrased as follows, by McMaster in his History of the People of the United States:

Why, it was asked, should a philosopher be made President? Is not the active, anxious, and responsible station of Executive illy suited to the calm, retired, and exploring tastes of a natural philosopher? Ability to impale butterflies and contrive turn-about chairs may entitle one to a college professorship, but it no more constitutes a claim to the Presidency than the genius of Cox, the great bridge-builder, or the feats of Ricketts, the famous equestrian. Do not the pages of history teem with evidences of the ignorance and mismanagement of philosophical politicians? John Locke was a philosopher, and framed a constitution for the colony of Georgia; but so full was it of whimsies that it had to be thrown aside. Condorcet, in 1793, made a constitution for France; but it contained more absurdities than were ever before piled up in a system of government, and was not even tried. Rittenhouse was another philosopher; but the only proof he gave of political talents was suffering himself to be wheedled into the presidency of the Democratic Society of Philadelphia. But, suppose that the title of philosopher is a good claim to the Presidency, what claim has Thomas Jefferson to the title of philosopher? Why, forsooth!

He has refuted Moses, disproved the story of the Deluge, made a penal code, drawn up a report on weights and measures, and speculated profoundly on the primary causes of the difference between the whites and blacks. Think of such a man as President! Think of a foreign minister surprising him in the act of anatomizing the kidneys and glands of an African to find out why the negro is black and odoriferous!

He has denied that shells found on the mountain tops are parts of the great flood. He has declared that if the contents of the whole atmosphere were water, the land would only be overflowed to the depth of fifty-two and one-half feet. He does not believe the Indians emigrated from Asia.

Every mail from the South brought accounts of rumblings and quakes in the Alleghanies and strange lights and blazing meteors in the sky. These disturbances in the natural world might have no connection with the troubles in the political world; nevertheless it was impossible not to compare them with the prodigies all writers of the day declare preceded the fatal Ides of March.

X

In New York, although a flourishing medical school had been in existence from 1769, there was an astonishing dearth of naturalists until about 1790. Governor Colden, the botanist and ethnologist, has died in 1776, and the principal medical men of the city, the Bards, Clossy, Jones, Middleton, Dyckman, and others confined their attention entirely to professional studies. A philosophical society was born in 1787, but died before it could speak. A society for the promotion of agriculture, arts, and manufactures, organized in 1791, was more successful, but not in the least scientific. Up to the end of the century New York State had but six men chosen to membership in the American Philosophical Society, and up to 1809 but five in the American Academy. Leaders, however, soon arose in Mitchill, Clinton, and Hosack.

Samuel Latham Mitchill, the son of a Quaker farmer [b. 1764, d. 1831], was educated in the medical schools of New York and Edinburgh, and in 1792 was appointed professor of chemistry, natural history, and philosophy in Columbia College. Although during most of his long life a medical professor and editor and for many years Representative and Senator in Congress, he continued active in the interests of general science. He made many contributions to systematic natural history, notably a History of the Fishes of New York, and his edition of Bewick's General History of Quadrupeds, published in New York in 1804 with notes and additions, and some figures of American animals, was the earliest American work of the kind. He was the first in America to lecture upon geology, and published several papers upon this science. His mineralogical exploration of the banks of the Hudson River in 1796, under the Society for the Promotion of Agriculture, Manufactures, and Useful Arts, founded by himself, was our earliest attempt at this kind of research, and in 1794 he published an essay on the Nomenclature of the New Chemistry, the first American paper on chemical philosophy, and engaged in a controversy with Priestley in defense of the nomenclature of Lavoisier, which he was the first American to adopt.

His discourse on The Botanical History of North and South America was also a pioneer effort. He was an early leader in ethnological inquiries and a vigorous writer on political topics. His Life of Tammany, the Indian chief (New York, 1795), is a classic, and he was well known to our grandfathers as the author of An Address to the Fredes or People of the United States, in which he proposed that Fredonia should be adopted as the name of the nation.

Doctor Mitchill was a poet[32] and a humorist and a member of the literary circles of his day. In The Croakers, Rodman Drake thus addressed him as The Surgeon-General of New York:

> It matters not how high or low it is
> Thou knowest each hill and vale of knowledge,
> Fellow of forty-nine societies
> And lecturer in Hosack's College.

Fitz-Greene Halleck also paid his compliments in the following terms:

> Time was when Doctor Mitchill's word was law,
> When Monkeys, Monsters, Whales and Esquimaux,
> Asked but a letter from his ready hand,
> To be the theme and wonder of the land.

These and other pleasantries, of which many are quoted in Fairchild's admirable History of the New York Academy of Sciences, gives us an idea of the provinciality of New York sixty years ago, when every citizen would seem to have known the principal local representatives of science, and to have felt a sense of personal proprietorship in him and in his projects.

Mitchill was a leader in the New York Historical Society, founder of the Literary and Philosophical Society, and of its successor, the Lyceum of Natural History, of which he was long president. He was also president of the New York Branch of the Linnæan Society of Paris, and of the New York State Medical Society, and surgeon-general of the State militia; a man of the widest influence and universally beloved. He served four terms in the House of Representatives, and was five years a member of the United States Senate.[33]

De Witt Clinton [b. 1769, d. 1828], statesman and philanthropist, United States Senator and governor of New York, was a man of similar tastes and capacities. What Benjamin Franklin was to Philadelphia in the middle of the eighteenth century, De Witt Clinton was to New York in the beginning of the nineteenth. He was the author of the Hibernicus Letters on the Natural History and Internal Resources of the State of New York (New York, 1822), a work of originality and merit. As president of the Literary and Philosophical Society he delivered, in 1814, an Introductory Discourse, which, like Barton's, in Philadelphia, ten years before, was productive of great good. It was, moreover, laden with the results of original and important observations in all departments of natural history. Another important paper was his Memoirs on the Antiquities of Western New York, printed in 1818.

Clinton's attention was devoted chiefly to public affairs, and especially to the organization of the admirable school system of New York, and other internal improvements. He did enough in science, however, to place him in the highest ranks of our early naturalists.[34]

Hosack has been referred to elsewhere as a pioneer in mineralogy and the founder of the first botanic garden. He was long president of the Historical Society, and exercised a commanding influence in every direction. His researches were, however, chiefly medical.

Samuel Akerly [b. 1785, d. 1845], the brother-in-law of Mitchill, a graduate of Columbia College, 1807, was an industrious worker in zoology and botany, and the author of the Geology of the Hudson River. John Griscom [b. 1774, d. 1852], one of the earliest teachers of chemistry, began in 1806 a career of great usefulness. "For thirty years," wrote Francis, "he was the acknowledged head of all other teachers of chemistry among us (in New York), and he kept pace with the flood of light which Davy, Murray, Gay Lussac, and Thenard and others shed on the progress of chemical philosophy at that day." About 1820 he went abroad to study scientific institutions, and his charming book, A Year in Europe, supplemented by his regular contributions to Silliman's Journal, commenting on scientific affairs in other countries, did much to stimulate the growth of scientific and educational institutions in America.

Francis tells us that he was for thirty years the acknowledged head of the teachers of chemistry in New York.[35]

A zealous promoter of zoology in those days was F. Adrian Vanderkemp, of Oldenbarnavelt, New York, who, in 1795, we are told, delivered an address before an agricultural society in Whitesburg, New York, in which he offered premiums for essays upon certain subjects, among which was one for the best anatomical and historical account of the moose, $50, or for bringing one in alive, $60.[36]

Having mentioned several American naturalists of foreign birth, it may not be out of place to refer to the American origin of an English zoologist of high repute, Doctor Thomas Horsfield, born in Philadelphia, in 1773, and after many years in the East became, in 1820, a resident of London, where he died in 1859. His name is prominent among those of the entomologists, botanists, and ornithologists of this century, especially in connection with Java.

XI

In New England science was more highly appreciated than in New York. Massachusetts had in John Adams a man who, like Franklin and Jefferson, realized that scientific institutions were the best protection for a democratic government, and to his efforts America owes its second scientific society—the American Academy of Arts and Sciences, founded in 1780. When Mr. Adams traveled from Boston to Philadelphia, in the days just before the Revolution, he several times visited at Norwalk, we are told, a curious collection of American birds and insects made by Mr. Arnold. This was afterwards sold to Sir Ashton Lever, in whose apartments in London Mr. Adams saw it again, and felt a new regret at our imperfect knowledge of the productions of the three kingdoms of nature in our land. In France his visits to the museums and other establishments, with the inquiries of academicians and other men of science and letters respecting this country and their encomiums on the Philosophical Society of Philadelphia, suggested to him the idea of engaging his native state to do something in the same good but neglected cause.[37]

The academy, from the first, was devoted chiefly to the physical sciences, and the papers in its memoirs for the most part relate to astronomy and meteorology.

Among its early members I find the names of but two naturalists: The Rev. Manasseh Cutler, pastor of Ipswich Hamlet, one of the earliest botanists of New England,[38] and William Dandridge Peck [b. 1763, d. 1882], the author of the first paper on systematic zoology ever published in America, a Description of Four Remarkable Fishes taken near the Piscataqua in New Hampshire, published in 1794.[39] Peck, after graduating at Harvard, lived at Kittery, New Hampshire, and first became interested in natural history by reading a wave-worn copy of Linnæus's System of Nature, which he obtained from the ship which was wrecked near his house. He became a good entomologist, and communicated much valuable material to Kirby in England, and was also one of our first writers on the fungi. He was the first to occupy the chair of natural history in Harvard University, to which he was appointed in 1800.

The Rev. Doctor Jedediah Morse [b. 1761; graduate of Yale, 1783; d. 1826] was the earliest of American geographers, and appears, especially in the later gazetteers published by him, to have printed important facts concerning the number and geographical distribution of the various Indian tribes.

The Connecticut Academy of Arts and Sciences was founded in 1799, one of the chief promoters being President Dwight [b. 1752, d. 1817], whose Travels in New England and New York, printed in 1821, abounds with scientific observations.

Another was E. C. Herrick [b. 1811, d. 1862], for many years librarian and subsequently treasurer of Yale College, whose observations upon the aurora, made in the latter years of the last century, are still frequently quoted; and later an active investigator of volcanic phenomena, and the author of a treatise on the Hessian fly and its parasites, the results of nine years' study; and of another on the existence of a planet between Mercury and the Sun.

Benjamin Silliman [b. in Trumbull, Connecticut, August 8, 1779; d. in New Haven, November 27, 1869], who, in 1802, became professor of chemistry at Yale, began there his career of usefulness as an organizer, teacher, and critic. One of his introductions to popular favor was the paper which he, in conjunction with Professor Kingsley, published, An account of the meteor which burst over Weston, in Connecticut, in December, 1807. This paper attracted attention everywhere, for the nature of meteors was not well understood in those days. Jefferson was reputed to have said in reference to it, that it was easier to believe that two Yankee professors could lie than to admit that stones could fall from heaven, but I think this must be pigeonholed with the millions of other slanders to which Jefferson was subjected in those days. I find in the papers by Rittenhouse and Madison, published twenty years before, by the Philosophical Society, matter-of-fact allusions to the falling of meteors to the earth.

Silliman was the earliest of American scientific lecturers who appeared before popular audiences, and, as founder and editor of the Journal of Science, did a service to science the value of which is beyond estimate or computation.

Benjamin Waterhouse, professor of the theory and practice of medicine in Harvard, 1783–1812, was one of the earliest teachers of natural history in America, and the author of a poem entitled The Botanist.[40]

The Rev. Jeremy Belknap [b. 1744, d. 1798], in his History of New Hampshire, and the Rev. Samuel Williams [b. 1743, d. 1817], in his Natural and Civil History of Vermont,[41] made contributions to local natural history, and Captain Jonathan Carver [b. 1732, d. 1780], in his Travels Through the Interior Parts of America, 1778, gave some meager information as to the zoology and botany of regions previously unknown.

In the South the prestige of colonial days seemed to have departed. Except Jefferson, the only naturalist in Virginia was Doctor James Greenway, of Dinwiddie County, a botanist of some merit. Mitchell returned to England before the Revolution, and Garden followed in 1784. H. B. Latrobe, of Baltimore, was an amateur ichthyologist, and Doctor James MacBride, of Pineville, South Carolina [b. 1784, d. 1817], was an active botanist. Doctor Lionel Chalmers [b. 1715, d. 1777], who was for many years the leader of scientific activity in South Carolina, was omitted in the previous address. A graduate of Edinburgh, he was for forty years a physician in Charleston. He recorded observations on meteorology from 1750 to 1760, the foundation of his Treatise on the Weather and Diseases of South Carolina [London, 1776], and published also valuable papers on pathology. He was the host and patron of many naturalists, such as the Bartrams.

There was no lack of men in the South who were capable of appreciating scientific work. Virginia had fourteen members in the American Philosophical Society from 1780 to 1800, while Massachusetts and New York had only six each, the Carolinas had eight, and Maryland six. The population of the South was, however, widely dispersed and no concentration of effort was possible. To this was due, no doubt, the speedy dissolution of the Academy of Arts and Sciences founded in Richmond in 1788.[42]

A name which should, perhaps, be mentioned in connection with this is that of Doctor William Charles Wells, whom it has been the fashion of late to claim as an American. It would be gratifying to be able to vindicate this claim, for Wells was a man of whom any nation might be proud. He was the originator of the generally accepted theory of the origin of dew, and was also, as Darwin has shown, the first to recognize and announce the theory of evolution by natural selection.[43] Unfortunately, Wells's science was not American science. We might with equal propriety claim as American the art of James Whistler, the politics of Parnell, the fiction of Alexandre Dumas, the essays of Grant Allen, or the science of Rumford and Le Vaillant.

Wells was the son of an English painter who emigrated in 1753 to South Carolina, where he remained until the time of the Revolution, when, with other loyalists, he returned to England. He was born during his father's residence in Charleston, but left the country in his minority; was educated at Edinburgh, and though he, as a young physician, spent four years in the United States, he was permanently established in London practice fully twenty-eight years before he read his famous letter before the Royal Society.

The first American naturalist who held definite views as to evolution was, undoubtedly, Rafinesque. In a letter to Doctor Torrey, December 1, 1832, he wrote:

The truth is that species, and perhaps genera also, are forming in organized beings by gradual deviations of shapes, forms, and organs taking place in the lapse of time. There is a tendency to deviation and mutation in plants and animals by gradual steps, at remote, irregular periods. This is a part of the great universal law of perpetual mutability in everything.

It is pleasant to remember that both Darwin and Wallace owed much of their insight into the processes of nature to their American explorations. It is also interesting to recall the closing lines, almost prophetic as they seem to-day, of the Epistle to the author of the Botanic Garden,[44] written in 1798 by Elihu Hubbard Smith, of New York, and prefixed to the American editions of The Botanic Garden:

Where Mississippi's turbid waters glide
And white Missouri pours its rapid tide;
Where vast Superior spreads its inland sea
And the pale tribes near icy empires sway;
Where now Alaska lifts its forests rude
And Nootka rolls her solitary flood.
Hence keen incitement prompt the prying mind
By treacherous fears, nor palsied nor confined;
Its curious search embrace the sea and shore
And mine and ocean, earth and air explore
Thus shall the years proceed—till growing time
Unfold the treasures of each different clime;
Till one vast brotherhood mankind unite
In equal bonds of knowledge and of right;
Thus the proud column, to the smiling skies
In simple majesty sublime shall rise,
O'er ignorance foiled, their triumph loud proclaim,
And bear inscribed, immortal, Darwin's name.

XII

During the three decades which made up the post-revolutionary period there were several beginnings which may not well be referred to in connection with individuals or localities.

The first book upon American insects was published in 1797, a sumptuously illustrated work, in two volumes, with 104 colored plates, entitled The Natural History of the rarer Lepidopterous Insects of Georgia. This was compiled by James E. Smith from the notes and drawings of John Abbot [b. about 1760], living in England in 1840, an accomplished collector and artist, who had been for several years a resident of Georgia, gathering insects for sale in Europe. Mr. Scudder characterizes him as the most prominent student of the life histories of insects we have ever had.[45]

There had, however, been creditable work previously done in what our entomologists are pleased to call the biological side of the science. As early as 1768, Colonel Landon Carter, of Sabine Hall, Virginia, prepared an elaborate paper, Observations concerning the Fly Weevil that Destroys the Wheat, which was printed by the American Philosophical Society,[46] accompanied by an extended report by The committee of husbandry. In the same year Moses Bartram presented his Observations on the Native Silkworms of North America.[47]

Organized effort in economic entomology appears to date from the year 1792, when the American Philosophical Society appointed a committee to collect materials for a natural history of the Hessian fly, at that time making frightful ravages in the wheat field, and so much dreaded in Great Britain that the import of wheat from the United States was forbidden by law. The Philosophical Society's committee was composed of Thomas Jefferson, at that time Secretary of State in President Washington's cabinet, Benjamin Smith Barton, James Hutchinson, and Caspar Wistar. In their report, which was accompanied by large drawings, the history of the little marauder was given in considerable detail.

The publication of Wilson's American Ornithology, beginning in 1808, was an event of great importance. It was in 1804 that the author, a schoolmaster near Philadelphia, decided upon his plan. In a letter to Lawson he wrote:

I am most earnestly bent on pursuing my plan of making a Collection of all the Birds of North America. Now, I don't want you to throw cold water on this notice, Quixotic as it may appear. I have been so long accustomed to the building of Airy Castles and brain Windmills that it has become one of my comforts of life, a sort of tough Bone, that amuses me when sated with the dull drudgery of Life.

I need not eulogize Wilson. Everyone knows how well he succeeded. He has had learned commentators and eloquent biographers. Our children pore over the narrative of the adventurous life of the weaver naturalist, and we all are sensible of the charms which his graceful pen has given to the life histories of the birds.

His poetical productions are immortal, and his lines to the Blue Bird and the Fisherman's Hymn are worthy to stand by the side of Bryant's Waterfowl, Trowbridge's Wood Pewee, Emerson's Titmouse, Thaxter's Sandpiper, and, possibly best of all, Walt Whitman's Mocking-Bird in Out of the Cradle endlessly Rocking.

Ichthyology in America dates also from these last years of the century. Garden was our only resident ichthyologist until Peck and Mitchill began their work, but Schœpf, the Hessian military surgeon, printed a paper on the Fishes of New York in 1787, and William Bryant, of New Jersey, and Henry Collins Flagg, of South Carolina, made observations upon the electric eel, in addition to those which Williamson, of North Carolina, laid before the Royal Society in 1775.

Paleontology had its beginnings at about the same time in the publication of Jefferson's paper on the Megalonyx or Great Claw in 1797.[48]

This early study of a fossil vertebrate was followed twenty years later by the first paper which touched upon invertebrates—that by Say on Fossil Zoology, in the first volume of Silliman's Journal. Lesueur seems to have brought from France some knowledge of the names of fossils and identified many species for the early American geologists.

Stratigraphical and physical geology also came in at this time, and will be referred to later.

The science of mineralogy was brought to American in its infancy. The first course of lectures upon this subject ever given in London was in the winter of 1793–94, by Schmeisser, a pupil of Werner. Doctor David Hosack, then a student of medicine at Edinburgh, was one of his hearers, and inspired by his enthusiasm began at once to form the collection of minerals which he brought to America on his return in 1794, which was the first mineralogical cabinet ever seen on this side of the Atlantic. This collection was exhibited for many years in New York (and in 1821 was given to Princeton College). Howard soon after obtained a select cabinet from Europe, and the museum of the American Philosophical Society acquired the Smith collection. In 1802 Mr. B. D. Perkins, a New York bookseller, brought from London a fine collection, which soon passed into the possession of Yale College, and in 1803 Doctor Archibald Bruce brought over one equally fine, which was made the basis of lectures when, in 1806, he became professor of mineralogy in Columbia College. George Gibbs, in 1805, imported the magnificent collection which was long in the custody of the American Geological Society. Seybert about the same time brought to Philadelphia the cabinet which, in 1813, was bought by the Academy of Natural Sciences and was lectured upon by Troost in 1814.

Much of the early botanical exploration was, however, carried out by European botanists: André Michaux [b. near Versailles, 1746; d. Madagascar, 1802], a pupil of the Jussieus and an experienced explorer, was sent by his Government, in 1785, to collect useful trees and shrubs for naturalization in France. He remained eleven years, made extensive explorations in the regions then accessible and as far west as the Mississippi, sent home immense numbers of living plants; and after his return, in 1796, published his treatise on the American Oaks,[49] and prepared the materials for his posthumous Flora Boreali-Americanas.

Francois André Michaux [b. near Versailles, 1770; d. at Vauréal, 1855] was his father's assistant in these early travels, and in 1802 and 1806 himself made botanical explorations in the Mississippi Valley. His botanical works were of great importance,[50] especially that known in its English translation as the North American Sylva, afterwards completed by Nuttall, and still the only work of the kind, though soon to be supplemented, we hope, by Professor Sargent's projected monographs.

Frederick Pursh [b. 1774, in Tobolsk, Siberia; d. June 11, 1820, in Montreal, Canada] carried on botanical explorations between 1799 and 1819, living from 1802 to 1805 in Philadelphia and from 1807 to 1810 in

New York. In 1814 he published in London his Flora Americæ Septen-trionalis. Pursh's Flora was largely based upon the labors of the American botanists Barton, Hosack, Le Conte, Peck, Clayton, Walter, and Lyon, and the botanical collection of Lewis and Clarke, and enumerated about 3,000 species of plants, while Michaux's, printed eleven years before, had only about half that number.

A. von Enslen collected plants at this time, in the South and West, for the Imperial Cabinet in Vienna. C. C. Robin, who traveled from 1802 to 1806 in what are now the Gulf States, wrote a botanical appendix to his Travels, published in 1807, on which Rafinesque founded his Florula Ludoviciana (New York, 1817).

Thaddeus Hænke [b. 1761, d. in Cochabamba, Bolivia, 1817] visited western North America with the Spaniards late in the last century, and made large collections of plants, which were sent to the National Museum of Bohemia, at Prague, and in part described in Presl's Reliquiæ Hænkianæ, 72 plates.

Archibald Menzies [b. 1754, d. 1842], an English naval surgeon, also collected on our Pacific Coast, under Vancouver, in 1780–1795, and his plants found their way to Edinburgh and Kew.

Captain Wangenheim, Surgeon Schoeph, of the Hessian contingent of the British army, Olaf Swartz, a Swedish botanical explorer, and others, also gathered plants in these early days, and in some instances published in Europe their botanical observations.

Other collectors of this same class were L. A. G. Bosc [1759–1828], who made botanical researches in the Carolinas during the last two years of the century, and returned to France in 1800 with a herbarium of 1,600 species. He also collected fishes, and his name is perpetuated in connection with at least two well known American fauna. Another was M. Milbert, who collected for Cuvier in New York, Canada, the Great Lake region, and the Mississippi Valley from 1817 to 1823.

The Baron Palisot de Beauvois [b. 1755, d. 1820] came from Santo

Domingo to America in 1791. He traveled extensively, and being a zoologist as well as a botanist, made observations upon our native animals,
particularly the reptiles.

It is to him that we owe the most carefully recorded of existing observations of young rattlesnakes crawling down their parent snakes' throats
for protection from enemies.

Most of these men did not contribute largely to the advancement of
American scientific institutes or affiliate with the naturalists of the day.

Of quite another type was the Count Luigi Castiglioni, who traveled,
soon after the Revolution, throughout the Eastern States and published
in 1790 two volumes of his travels.[51]

The Count Volney [b. at Craon February 3, 1757, d. in Paris April 25,
1820], traveler, salesman, and historian, traveled in this country from
1795 to 1798, and in 1803, while a senator of the French Republic,
published his famous work upon the United States, containing his observations upon its soil and its climate, and upon the Indians, together with
the first doctrines of the language of the Miamis,[52] and also giving a
description of the physical and botanical features of the country. Volney
was an admirer and intimate friend of Franklin, and it was in his home at
Passy, we are told, that he conceived the idea of his most famous book,
Les Ruines.[53]

Among the traditions of Fauquier County, Virginia, is one which is of
interest to naturalists, since it relates to an incident showing the interest
of our first President in science:

About the year 1796 [runs the story], at the close of a long summer's day, a
stranger entered the village of Warrenton. He was alone and on foot, and his
appearance was anything but prepossessing. His garments, coarse and dust
covered, indicated an individual in the humble walks. From a cane across his
shoulders was suspended a handkerchief containing his clothing. Stopping in

front of Turner's tavern, he took from his hat a paper and handed it to a gentlemen standing on the steps; it read as follows:

The celebrated historian and naturalist

Volney needs no recommendation from

G. WASHINGTON.

In 1801 Jefferson began his eight years of Presidency. Since he was the only man of science who has ever occupied the Chief Magistracy, he has a right to a high place in the esteem of such a society as ours, and I only regret that, having spoken of him at length a year ago, I can not now discuss his scientific career in all its aspects.

I then spoke of the credit which was due to him for beginning so early as 1780 to agitate the idea of a Government exploring expedition to the Pacific, which culminated in the sending out by Congress of the expedition of Lewis and Clarke, in 1803. Captain Lewis [b. 1774, d. 1809], the leader of this expedition, was a young Virginian, the neighbor and for some years the private secretary of President Jefferson. He set out in the summer of 1803, accompanied by his associate, Captain Clarke, and twenty-eight men. They entered the Missouri May 14, 1804, before the middle of the following July had reached the great falls, and by October were upon the western slope, where, embarking in canoes upon the Kouskousky, a branch of the Columbia, they descended to its mouth, where they arrived on the 15th of November, 1805. The following spring they retraced their course, arriving in St. Louis in September.[54] The results of the expedition were first made known in Jefferson's message to Congress read February 19, 1806.

Doctor Asa Gray, in a recent letter, says:

I have reason to think that Michaux suggested to Jefferson the expedition which the latter was active in sending over to the Pacific. I wonder if he put off Michaux for the sake of having it in American hands?[55]

The idea of an expedition to the Pacific was one which was likely to occur to any thoughtful American, and was, after all, simply the continuance of a plan as old as the Spanish days of discovery. Jefferson, at all events, was an active promoter of all such enterprises, and after a quarter of a century's effort the expedition was dispatched, while in 1805 General Z. M. Pike was sent to explore the sources of the Mississippi River and the western parts of Louisiana, penetrating as far west as Pike's Peak, a name which still remains as a memento of this enterprise.

The organization of these early expeditions marked the beginning of one of the most important portions of the scientific work of our Government—the investigation of the resources and natural history of the public domain. The expeditions of Lewis and Clarke and of Pike were the precursors and prototypes of the magnificent organization now accomplishing so much for science under the charge of Major J. W. Powell.

As early as 1806 Jefferson, inspired by Patterson and Hassler, urged the establishment of a national coast survey, and in this was earnestly supported by his Secretary of the Treasury, Albert Gallatin, who drew up a learned and elaborate project for its organization, and an act authorizing its establishment was passed in 1807. During his Administration, in 1802 the first scientific school in this country was established—the Military Academy at West Point. The Military Academy was a favorite project of General Washington, who is said to have justified his anxiety for its establishment by the remark that "An army of asses led by a lion is vastly superior to an army of lions led by an ass."

Jefferson has been heartily abused for not gratifying Alexander Wilson's request to be appointed naturalist to Pike's expeditions. It is possible that even in those days administrators were hampered by lack of financial resources. It must also be remembered that in 1804 Wilson was simply an enthusiastic projector of ornithological undertakings, and had done nothing whatever to establish his reputation as an investigator.

One of Jefferson's first official acts was to throw his Presidential mantle over Priestley. Two weeks after he became President of the United States he wrote these words:

It is with heartfelt satisfaction that, in the first moments of my public action, I can hail you with welcome to our land, tender to you the homage of its respect and esteem, cover you under the protection of those laws which were made for the wise and good like you, and disclaim the legitimacy of that libel on legislation which, under the form of a law, was for sometime placed among them.

. . . Yours is one of the few lives precious to mankind, and for the continuance of which every thinking man is solicitous. Bigots may be an exception. What an effort, my dear sir, of bigotry in politics and religion have we gone through. . . . All advances in science were prescribed as innovations. They pretended to praise and encourage education, but it was to be the education of our ancestors. We were to look backwards, not forwards for improvement; the President [Washington] himself declaring in one of his answers to addresses that we were never to expect to go beyond them in real science. This was the real ground of all the attacks on you; those who live by mystery and *charlatanerie* fearing you would render them useless by simplifying the Christian philosophy, the most sublime and benevolent but most perverted system that ever shone on man, endeavored to crush your well-earned and well-deserved fame.[56]

XIII

With the close of the first decade ended the first third of a century since the Declaration of Independence. We have now passed in review a considerable number of illustrious names and have noted the inception of many worthy undertakings.

"Still, however," in the words of Silliman, "although individuals were enlightened, no serious impression was produced on the public mind; a few lights were indeed held out but they were lights twinkling in an almost impervious gloom.[57]

This was a state of affairs not peculiar to America. A gloom no less oppressive had long obscured the intellectual atmosphere of the Old World. There were a goodly number of men of science, and many important discoveries were being made, but no bonds had yet been formed to connect the interests of the men of science and the men of affairs.

Speculative science, in the nature of things, can only interest and attract scholarly men, and though its results, concisely and attractively stated, may have a passing interest to a certain portion of every community, it is only by its practical applications that it secures the hearty support of the community at large.

Huxley, in his recent discourse upon The Advance of Science in the last Half Century,[58] has touched upon this subject in a most suggestive and instructive manner, and has shown that Bacon, with all his wisdom, exerted little direct beneficial influence upon the advancement of natural knowledge, which has after all been chiefly forwarded by men like Galileo and Harvey, Boyle and Newton, "who would have done their work quite as well if neither Bacon nor Descartes had ever propounded their views respecting the manner in which scientific investigation should be pursued."

I think we should look upon Bacon as the prophet of modern scientific thought, rather than its founder. It is no doubt true, as Huxley has said, that his "scientific insight" was not sufficient to enable him to shape the future course of scientific philosophy, but it is scarcely true that he attached any undue value to the practical advantages which the world as a whole and incidentally science itself were to reap from the applications of scientific methods to the investigation of nature.

Even though the investigations of Descartes, Newton, Leibnitz, Boyle, Torricelli, and Malpighi had directly helped no man to either wealth or comfort, the cumulative results of their labors, and those of their pupils and associates, resulted in a condition of scientific knowledge from which, sooner or later, utilitarian results must necessarily have sprung.

It is true, as Huxley tells us, that at the beginning of this century weaving and spinning were still carried on with the old appliances; true that nobody could travel faster by sea or by land than at any previous time in the world's history, and true that King George could send a message from London to York no faster than King John might have done. Metals were still worked from their ores by immemorial rule of thumb, and the center of the iron trade of these islands was among the oak forests of Sussex, while the utmost skill of the British mechanics did not get beyond the production of a coarse watch.

It can not be denied that although the middle of the eighteenth century was illuminated by a host of great names in science, chemists—biologists, geologists—English, French, German, and Italian, the deepening and broadening of natural knowledge had produced next to no immediate practical benefits. Still I can not believe that Bacon, the prophet, would have been so devoid of "scientific insight" as to have failed to foresee at this time the ultimate results of all this intellectual activity.

But Huxley says:

Even if, at this time, Francis Bacon could have returned to the scene of his greatness and of his littleness, he must have regarded the philosophic world which praised and disregarded his precepts with great disfavor. If ghosts are consistent he would have said, "these people are all wasting their time, just as Gilbert and Kepler and Galileo and my worthy physician Harvey did in my day. Where are the fruits of the restoration of science which I promised? This accumulation of bare knowledge is all very well, but *cui bono*? Not one of these people is doing what I told him specially to do, and seeking that secret of the cause of forms which will enable him to deal at will with matter and superinduce new nature upon old foundations."

As Huxley, however, proceeds himself to show in the discussion which immediately follows this passage, a "new nature, begotten by science upon fact," has been born within the past few decades, and, pressing itself daily and hourly upon our attention, has worked miracles which have not only modified the whole future of the lives of mankind, but has reacted constantly upon the progress of science itself.

It is to the development of this new nature, then in its very infancy, that we must look for the revival of interest in science on this side of the Atlantic.

The second decade of the century was marked by a great accession of interest in the sciences. The second war with Great Britain having ended, the country, for the first time since colonial days, became sufficiently tranquil for peaceful attention to literature and philosophy. The end of the Napoleonic wars and the restoration of tranquillity to Europe tended to scientific advances on the other side of the Atlantic, and the results of the labors of Cuvier, whose glory was now approaching its zenith, of Brongniart, of Blainville, of Jussieu, of Decandolle, of Werner, of Hutton, of Buckland, of De la Beche, of Magendie, of Humboldt, Daubuisson, Berzelius, Von Buch, or Heschel, of Laplace, of Young, of Fresnel, of Oersted, of Cavendish, of Lavoisier, Wollaston, Davy, and Sir William Hooker, were eagerly welcomed by hundreds in America.

"In truth," wrote one who was among the most active in promoting these tendencies—"a thirst for the natural sciences seemed already to pervade the United States like the progress of an epidemic."

The author of these enthusiastic words was Amos Eaton [b. in Chatham, New York, 1776, d. May 6, 1842], one of the most interesting men of his day. In 1816, at the age of forty, he abandoned the practice of law and went to New Haven to attend Silliman's lectures on mineralogy and geology. He was a man of great force and untiring energy, and one of the pioneers of American geology; though the name, Father of American Geology, sometimes applied to him, would seem to belong more appropriately to Maclure, or, perhaps, to Mitchill. He was, however, only some eight years later than Maclure in beginning geological field work. Eaton's *Index to the Geology of the Northern States of America*, printed in 1817, was the first strictly American treatise, and seems to have had a very stimulating effect. He was preeminently an agitator and an educator. He traveled many thousands of miles on foot throughout New England and New York, delivering in the meantime, at the principal towns, short courses of lectures on natural history. In March, 1817, having received an invitation to aid in the introduction of the natural sciences in Williams College, his Alma Mater, he delivered a course of lectures in Williams-town. "Such," he remarks, "was the zeal at this institution, that an un-controlable enthusiasm for Natural History, took possession of every mind; and other departments of learning were, for a time, crowded out of College. The College authorities allowed twelve students each day (72 per week) to devote their whole time to the collection of minerals, plants, etc., in lieu of all other exercises."[59]

In April, 1818, he went to Albany, on the special invitation of Gover-nor De Witt Clinton, and delivered a course of lectures on natural his-tory. "In Albany I found," wrote he, "Doctor T. Romeyn Beck, and in Troy, Doctors Burrett, Robbins, and Dale, zealous beyond description in the cause of natural science. By the exertions of these gentlemen a taste for the study of nature was strongly excited in those two cities, especially for that of geology. They, together with several others, had become mem-bers of the New York Lyceum of Natural History, and in the fall of 1818 established a society of the same name and upon a similar plan in Troy. Collections were made with such zeal that in the course of a few months Troy could boast of a more extensive collection of American geological specimens than Yale College or any other institution upon this conti-nent."[60]

"In this period," remarked Bache, "the prosecution of mathematics and physical science was neglected; indeed, barely kept alive by the calls for boundary and land surveys of the more extended class, by the exertions necessary in the lecture room, or by isolated volunteer efforts.

"As the country was explored and settled the unworked mine of natural history was laid open, and the attention of almost all the cultivators of science was turned toward the development of its riches.

"Descriptive natural history is the pursuit which emphatically marks that period. As its exponent, may be taken the admirable descriptive mineralogy of Cleaveland, which seemed to fill the measures of that day and be, as it were, its chief embodiment, appearing just as the era was passing away."[61]

The leading spirits of the day seem to have been Silliman, Hare, Maclure, Mitchill, Gibbs, Cleaveland, De Witt Clinton, and Caspar Wistar.

Names familiar to us of the present generation began now to appear in scientific literature. Isaac Lea began to print his memoirs on the Unionidae; Edward Hitchcock, principal of the Deerfield Academy, was writing his first papers on the geology of Massachusetts; Professor Chester Dewey, of Williams College [b. 1781, d. 1867], afterwards known to us all from his excellent work upon the Carices, was discussing the mineralogy and geology of Massachusetts; Doctor John Torrey, also to be famous as a botanist, was then devoting his attention to mineralogy and chemistry; Doctor Jacob Porter was making botanical observations in central Massachusetts; quaint old Caleb Atwater, at that time almost the only scientific observer west of the Alleghenies, was discussing the origin of prairies, meteorology, botany, geology, mineralogy, and scenery of the Ohio country, and a little later the remains of mammoths.

Professor J. W. Webster, of Boston, was making general studies in geology; the Rev. Elias Cornelius and Mr. John Grammer were writing of the geology of Virginia; Mr. J. A. Kain, upon that of Tennessee, I. P. Brace, that of Connecticut, and James Pierce, that of New Jersey.

To this period belonged the brilliant Constantine Rafinesque, with Torrey, Silliman, Cleaveland, Gibbs, James, Schoolcraft, Gage, Akerly, Mitchill, Dana, Beck, and Featherstonhaugh.

Doctor Henry R. Schoolcraft, afterwards prominent in ethnology, printed, in 1819, his View of the Lead Mines of Missouri, the first from American contributors to economic geology; and in the same year his Transallegania, a mineralogical poem, probably the last as well as the first of its kind written in America. In 1821 he published a scholarly Account of the Native Copper on the Southern Shore of Lake Superior.[62]

Mineralogy and geology were the most popular of the sciences.

American geology dated its beginning from this previous decade. Professor S. L. Mitchill was one of the first to call attention to the teachings of Kirwan and the pioneers of European geology, and very early in the century began to instruct the students of Columbia College in the principles of geology as then understood. He published Observations on the Geology of America, and also edited a New York edition of Cuvier's History of the Earth, contributing to this work an appendix which was constantly quoted by early writers.

The first geological explorer was William Maclure [b. in Ayr, Scotland, 1763; d. in San Angel, Mexico, March 23, 1840], a Scotch merchant who amassed a large fortune by commercial connections with this country, and became a citizen of the United States about 1796. His most important service to American science was that of a patron, for he was a liberal supporter of the infant Academy of Sciences in Philadelphia, and for twenty-two years its president, besides being an upholder of other important enterprises.

The publication, in 1809, of his Observations on the Geology of the United States marks the beginning of American geographical geology and the first attempt at a geological survey of the United States. This had long been the subject of his ambition, and in order to prepare himself for the task he had spent several years in travel throughout Europe, making observations and collecting objects in natural history, which he forwarded to the country of his adoption.

His undertaking was undoubtedly a remarkable one. "He went forth with his hammer in his hand and his wallet on his shoulder, pursuing his researches in every direction, visiting almost every State and Territory, wandering often amidst pathless tracts and dreary solitudes until he had crossed and recrossed the Allegheny Mountains not less than fifty times. He encountered all the privations of hunger, thirst, fatigue, and exposure, month after month and year after year, until his indomitable spirit had conquered every difficulty, and crowned his enterprise with success,"[63] and after the publication of his memoir he devoted eight years more to collecting materials for a second and revised edition.

The geological map of the United States, published in 1809, appears to have been the first of its kind ever attempted for an entire country. Smith's geological map of England was six years later, and Greenough's still subsequent in date.

The publication in London in 1813 of Bakewell's Introduction to Geology seems to have given a great stimulus to geological researches in this country, as may be judged from the publication of an American edition a year or two later.

Mitchill, Bruce, and Maclure soon had a goodly band of associates. Naturalists were not confined to limited specialities in those days, and we find all the chemists, botanists, and zoologists absorbed in the consideration of geological problems. Maclure and most of the Americans were disciples of Werner.

Silliman, writing in 1818, said:

A grand outline has recently been drawn by Mr. Maclure with a masterly hand and with a vast extent of personal observation and labor; but, to fill up the detail, both observation and labor still more extensive are demanded; nor can the object be effected till more good geologists are formed and distributed over our extensive territory.

On the 6th of September, 1819, the American Geological Society was organized in the philosophical room of Yale College, an event of great importance in the history of science, hastening, as it seems to have done, the establishment of State surveys and stimulating observation throughout the country. This society, which continued in existence until about 1826, may fairly be considered the nucleus of the Association of American Geologists and Naturalists, and, consequently, of the American Association for the Advancement of Science. Members appended to their names the symbols M. A. G. S., and it was for a time the most active of American scientific societies.

The characteristics of the leading spirits were summed up by Eaton at the time of its beginning:

The president of the American Geological Society, William M'Clure, has already struck out the grand outline of North American geographical geology. The first vice-president, Col. G. Gibbs, has collected more facts and amassed more geological and mineralogical specimens than any other individual of the age. The second vice-president, Professor Silliman, gives the true scientific dress to all the naked mineralogical subjects which are furnished to his hand. The third vice-president, Professor Cleaveland, is successfully employed in elucidating and familiarizing those interesting sciences; and thus smoothing the rugged paths of the student. Professor Mitchill has amassed a large store of materials, and annexed them to the labors of Cuvier and Jameson. But the drudgery of climbing cliffs and descending into fissures and caverns, and of traversing in all directions our most rugged mountainous districts, to ascertain the distinctive characters, number, and order of our strata, has devolved on me.[64]

Eaton has very fairly defined his own position among the early geologists, which was that of an explorer and pioneer. The epithet, Father of American Geology, which has sometimes been applied to him, might more justly be bestowed upon Maclure, or even upon Mitchill. The name of Amos Eaton [b. 1776, d. 1872] will always be memorable, on account of his connection with the geological survey of New York, which was begun in 1820, at the private expense of Hon. Stephen Van Rensselaer; also as the founder, in 1824, of the Rensselaer Polytechnic Institute, the first of its class on the continent.

The State of New York was not preeminently prompt in establishing an official survey, but the liberality of Van Rensselaer and the energy of Eaton gave to New York the honor of attaching the names of its towns and counties to a large number of the geological formations of North America.

In these early surveys Eaton was associated with Doctor Theodore Romeyn Beck and Mr. H. Webster, naturalist and collector, one of the first being a survey of the county of Albany, under the special direction of a County Agricultural Society, followed by similar surveys of Rensselaer County and Saratoga County and others along the Erie Canal.

In July, 1818, Professor Silliman began the publication of the American Journal of Science, which has been for more than two-thirds of a century the most prominent register of the scientific progress of this continent. Silliman's journal succeeded, and far more than replaced, the American Mineralogical Journal, the earliest of American scientific periodicals, which was established in New York in 1810 by Doctor Archibald Bruce, and which was discontinued after the close of the first volume, in 1814, on account of the illness and untimely death of its projector.[65] The Mineralogical Journal was not so limited in scope as in name, and was for a time the principal organ of our scientific specialists.

We can but admire the spirit of Silliman, who remarks in the preface to the third volume:

It must require several years from the commencement of the work to decide the question [whether it is to be supported], and the editor (if God continues his life and health) will endeavor to prove himself neither impatient nor querulous during the time that his countrymen hold the question undecided, *whether there shall be an American Journal of Science and Arts.*

In the fall of 1822 he announced that a trial of four years had decided the point that the American public would support this journal.

Prior to the establishing of Silliman's journal, the principal organs of American science were the Medical Repository, commenced in 1798, of which Doctor Mitchill was the chief proprietor; the New York Medical and Physical Journal, conducted chiefly by Doctor Hosack; the Boston Journal of Philosophy and the Arts, and other similar periodicals. Our students looked chiefly, however, to the English journals: Tilloch's Philosophical Magazine and Nicholson's Journal of Natural Philosophy, and later, Thomson's Annals of Philosophy, the Annales de Chimie.

The American Monthly Magazine, established in 1814 by Charles Brockden Brown, was fully as much devoted to science as to literature, and an examination of this and other journals of the early portion of the century will, I think, satisfy the student that scientific subjects were more seriously considered by our ancestors than by the Americans of to-day. The American Monthly published elaborate reviews of technical works, such as Cleaveland's Mineralogy, and summaries of the world's progress in science, as well as the monthly proceedings of all the scientific societies in New York, and papers on systematic zoology and botany by Rafinesque.

In 1812 the American Antiquarian Society was established at Worcester, and before 1820, when its first volume of Transactions appeared, had collected 6,000 books and "a respectable cabinet." This was a pioneer effort in ethnological science. Archæologia Americana contained papers by Mitchill, Atwater, and others, chiefly relating to the aboriginal population of America. The name of Isaiah Thomas, LL.D. [b. in Boston 1749, d. in Worcester 1831], the founder and first president of the society, who at his own expense erected a building for its accommodation and endowed its first researches, should be remembered with gratitude by American naturalists. He was one of the most eminent of American printers, and was styled by DeWarville The Didot of America.

In 1812 the Academy of Natural Sciences of Philadelphia was founded, the outgrowth of a social club whose members, we are told, had no conception of the importance of the work they were undertaking when, in a spirit of burlesque, they assumed the title of an academy of learning.

In 1816 the Coast Survey, after years of discussion, was placed in action under the supervision of Hassler (who had been appointed its head as early as 1811), but two years later, the work going on too slowly to please the Government, it was stopped.

The Linnæan Society of New England, established in Boston about this time, was the precursor of the Boston Society of Natural Science.

The publication of an American edition of Rees's Cyclopædia, in Philadelphia, was begun in 1810, and the forty-seventh volume completed in 1824. This was an event in the history of American science, for it furnished employment and thus fostered the investigations of several eminent naturalists, among whom were Alexander Wilson, Thomas Say, and Ord, while at the same time it fostered a taste for science in the United States and gave currency to several rather epoch-making articles, such as Say's upon Conchology and Entomology.

Mr. Bradbury, the publisher of this cyclopædia, was the first of a goodly company of liberal and far-seeing publishers, who have done much for science in this country by their patronage of important scientific publications.

In 1817 Josiah Meigs, Commissioner of the Land Office, issued a circular to the several registers of the land offices of the United States requiring them to keep daily meteorological observations, and also to report upon such phenomena as the times of the unfolding of leaves of plants and the dates of flowering, the migrations of birds and fishes, the dates of spawning of fishes, the hibernation of animals, the history of locusts and other insects in large numbers, the falling of stones and other bodies from the atmosphere, the direction of meteors, and discoveries relative to the antiquities of the country.

It does not appear that anything ever resulted from this step, but it is referred to as an indication that, seventy years ago, our Government was willing to use its civil-service officials in the interest of science. A few years later the same idea was carried into effect by the Smithsonian Institution.

In those early days each of the principal cities had public museums founded and supported by private enterprise. Their proprietors were men of scientific tastes, who affiliated with the naturalists of the day and placed their collections freely at the disposal of investigators.

The earliest was the Philadelphia Museum, established by Charles Willson Peale, and for a time housed in the building of the American Philosophical Society. In 1800 it was full of popular attractions.

There were a mammoth's tooth from the Ohio, and a woman's shoe from Canton; nests of the kind used to make soup of, and a Chinese fan 6 feet long; bits of asbestus, belts of wampum, stuffed birds and feathers from the Friendly Islands, scalps, tomahawks, and long lines of portraits of great men of the Revolutionary war. To visit the Museum, to wander through the rooms, play upon the organ, examine the rude electrical machine, and have a profile drawn by the physiognomitian, were pleasures from which no stranger to the city ever refrained.

Doctor Hare's oxyhydrogen blowpipe was shown in this museum by Mr. Rubens Peale as early as 1810.

The Baltimore Museum was managed by Rembrandt Peale, and was in existence as early as 1815 and as late as 1830.[66]

Earlier efforts were made, however, in Philadelphia. Doctor Chovet, of that city, had a collection of wax anatomical models made by him in Europe, and Professor John Morgan, of the University of Pennsylvania, who learned his methods from the Hunters in London and Sué in Paris, was also forming such a collection before the Revolution.[67]

The Columbian Museum and Turrell's Museum, in Boston, are spoken of in the annals of the day, and there was a small collection in the attic of the State House in Hartford.

The Western Museum, in Cincinnati, was founded about 1815, by
Robert Best, M.D., afterwards of Lexington, Kentucky, who seems to
have been a capable collector, and who contributed matter to Godman's
American Natural History. In 1818 a society styled the Western Museum
Society was organized among the citizens, which, though scarcely a
scientific organization, seems to have taken a somewhat liberal and
public-spirited view of what a museum should be. To the naturalists of
to-day there is something refreshing in such simple appeals as the following:

In collecting the fishes and reptiles of the Ohio the managers will need all the
aid which their fellow-citizens may feel disposed to give them. Although not a
very interesting department of zoology, no object of the society offers so great
a prospect of novelty as that which embraces these animals.

The obscure and neglected race of insects will not be overlooked, and any
specimen sufficiently perfect to be introduced into a cabinet of entomology
will be thankfully received.[68]

Major John Eatton LeConte, U.S.A. [b. 1784, d. 1860], was a very
successful student of botany and zoology. He published many botanical
papers and contributions to descriptive zoology, and also in Paris, in conjunction with Boisduval, the first installment of a work, of which he was
really sole author, upon the Lepidoptera of North America.[69]

The elder brother, Doctor Lewis LeConte [b. 1782, d. 1838], was
equally eminent as an observer, and was for forty years one of the most
prominent naturalists in the South. On his plantation in Liberty County,
Georgia, he established a botanical garden and a chemical laboratory. His
zoological manuscripts were destroyed in the burning of Columbia just
at the close of the civil war, but his observations, which he was averse to
publishing in his own name, were, we are told, embodied in the writings
of his brother, of Stephen Elliott, of the Scotch botanist, Gordon,[70] of
Doctor William Baldwin and others.[71,72]

Stephen Elliott, of Charleston, South Carolina [b. 1711, d. 1830], was a graduate of Yale in the class of 1791 and, while prominent in the political and financial circles of his State, found time to cultivate science. He founded in 1813 the Literary and Philosophical Society of South Carolina, and was its first president; and in 1829 was elected professor of natural history and botany in the South Carolina Medical College, which he aided to establish. He published the Botany of South Carolina and Georgia (Charleston, 1821–1827), having been assisted in its preparation by Doctor James McBride; and had an extensive museum of his own gathering. The Elliott Society of Natural History, founded in 1853, or before, and subsequently continued under the name of the Elliott Society of Science and Art, 1859–1875, was named in memory of this public-spirited man.

Jacob Green [b. 1790, d. 1841], at different times professor in the College of New Jersey and in Jefferson Medical College, was one of the old school naturalists, equally at home in all of the sciences. His paper on Trilobites (1832) was our first formal contribution to invertebrate paleontology; his Account of some new Species of Salamanders,[73] one of the earliest steps in American herpetology; his Remarks on the Unios of the United States,[74] the beginning of studies subsequently extensively prosecuted by Lea and some other entomologists. He also wrote upon the crystallization of snow, and was the author of Chemical Philosophy, Astronomical Researches, and a work upon Botany of the United States.

The earlier volumes of Silliman's Journal were filled with notes of his observations in all departments of natural history.

José Francisco Correa da Serra, secretary of the Royal Academy of Lisbon, was resident in Philadelphia in 1813, in the capacity of Portuguese minister, and affiliated with our men of science in botanical and geological interests. In 1814 he lectured on botany in the place of B. S. Barton, and also published several botanical papers, as well as one upon the soil of Kentucky.

Alire Raffenau Delile, formerly a member of Napoleon's scientific expedition to Egypt and the editor of the Flora of Egypt, was in New York about this time, for the purpose of completing his medical education, and seems to have done much to stimulate interest in botanical studies.

To this as well as to the subsequent period belonged Doctor Gerard Troost [b. in Holland, March 15, 1776; educated in Leyden; d. in Nashville, August 17, 1850], a naturalist of Dutch birth and education, who came to Philadelphia in 1810, and was a founder and the first president of the Philadelphia Academy. In 1826 he founded a geological survey of the environs of Philadelphia; in 1827 became professor of chemistry, mineralogy, and geology in the University of Nashville. As State geologist of Tennessee from 1831 to 1849 he published some of the earliest State geological reports.

Another expedition well worthy of mention, though not exceedingly fruitful, was one made under the direction of Mr. Maclure, president of the Philadelphia Academy, to the sea islands of Georgia and the Florida peninsula. The party consisted of Maclure, Say, Ord, and Titian R. Peale, and the results, though not embodied in a formal report, may be detected in the scientific literature of the succeeding years. This was early in 1818, while Florida was still under the dominion of Spain, and the expedition was finally abandoned, owing to the hostile attitude of the Seminole Indians in that territory.

XIV

The third decade of the century, beginning with 1820, was marked by a continuation of the activities of that which preceded. In 1826 there were in existence twenty-five scientific societies, more than half of them especially devoted to natural history[75] and nearly all of very recent origin.

The leading spirits were Mitchill, Maclure, Webster, Torrey, Silliman, Gibbs, LeConte, Dewey, Hare, Hitchcock, Olmsted, Eliot, and T. R. Beck.

Nathaniel Bowditch [b. 1773, d. 1838], in 1829, began the publication of his magnificent translation of the Mécanique Celeste of La Place, with those scholarly commentations which secured him so lofty a place among the mathematicians of the world.

Still more important was the lesson of his noble devotion of his life and fortune to science. The greater part of his monumental work was completed, we are told, in 1817, but he found that to print it would cost $12,000, a sum far beyond his means. A few years later, however, he began its publication from his own limited means, and the work was continued, after his death, by his wife. The dedication is to his wife, and tells us that "without her approbation the work would not have been undertaken."

Another person was W. C. Redfield [b. 1789, d. 1857], who, in 1827, promulgated the essential portions of the theory of storms, which is now pretty generally accepted, and which was subsequently extended by Sir William Reid in Barbados and Bermuda, and greatly modified by Professor Loomis, of New Haven. An eloquent eulogy of Redfield was pronounced by Professor Denison Olmsted at the Montreal meeting of the American Association in 1857.[76]

Among the rising young investigators of this period were Joseph Henry, A. D. Bache, C. U. Shepard, the younger Silliman, Henry Seybert, William Mather, Ebenezer Emmons, Percival, the poet geologist, DeKay, Godman, and Harlan.

The organization, in 1824, of the Rensselaer School, afterwards the Rensselaer Polytechnic Institute, at Troy, marked the beginning of a new era in scientific and technological education. Its principal professors were Amos Eaton and Doctor Lewis C. Beck.

In 1820 an expedition was sent by the General Government to explore the Northwestern Territory, especially the region around the Great Lakes and the sources of the Mississippi. This was under charge of General Lewis Cass, at that time governor of Michigan Territory. Henry R. Schoolcraft accompanied this expedition as mineralogist, and Captain D.

B. Douglass, U. S. A., as topographical engineer; and both of these sent home considerable collections reported upon by the specialists of the day. Cass himself, though better known as a statesman, was a man of scientific tastes and ability, and his Inquiries respecting the History, Traditions, Languages, etc., of the Indians, published at Detroit in 1823, is a work of high merit.

Long's expeditions into the far West were also in progress at this time, under the direction of the General Government; the first, or Rocky Mountain, exploration in 1819–20; the second to the sources of the St. Peter's, in 1823. In the first expedition Major Long was accompanied by Edwin James as botanist and geologist, who also wrote the Narrative published in 1823. The second expedition was accompanied by William H. Keating, professor of mineralogy and chemistry in the University of Pennsylvania, who was its geologist and historiographer. Say was the zoologist of both explorations. De Schweinitz worked up the botanical material which he collected.

The English expeditions sent to Arctic North America under the command of Sir John Franklin were also out during these years, the first from 1819 to 1822, the second from 1825 to 1827, and yielded many important results. To naturalists they have an especial interest, because Sir John Richardson, who accompanied Franklin as surgeon and naturalist, was one of the most eminent and successful zoological explorers of the century, and had more to do with the development of our natural history than any other man not an American.

His natural history papers in Franklin's reports, 1823 and 1828, his Fauna Boreali Americana, published between 1827 and 1836, his report upon the Zoology of North America, are all among the classics of our zoological literature.[77]

The third decade was somewhat marked by a renewal of interest in zoology and botany, which had, during the few preceding years, been rather overshadowed by geology and mineralogy.

Rafinesque had retired to Kentucky, where, from his professor's chair in Transylvania University, he was issuing his Annals of Nature and his Western Minerva; and his brilliancy being dimmed by distance, other students of animals had a chance to work.

One of the most noteworthy of the workers was Thomas Say [b. 1787, d. 1834], who was a pioneer in several departments of systematic zoology. A kinsman of the Bartram's, he spent many of his boyhood days in the old botanic garden at Kingsessing, in company with the old naturalist, William Bartram, and the ornithologist Wilson. At the age of twenty-five, having been unsuccessful as an apothecary, he gave his whole time to zoology. He slept in the hall of the Academy of Natural Sciences, where he made his bed beneath the skeleton of a horse, and fed himself upon bread and milk. He was wont, we are told, to regard eating as an inconvenient interruption to scientific pursuits, and to wish that he had been created with a hole in his side through which his food might be introduced into his system. He built up the museum of the society, and made extensive contributions to biological science.

His article on conchology, published in 1816 in the American edition of Nicholson's Cyclopædia, was the foundation of that science in this country, and was republished in Philadelphia in 1819, with the title, A Description of the land and fresh-water Shells of the United States.

This work [remarked a contemporary] ought to be in the possession of every American lover of natural science. It has been quoted by M. Lamarck and adopted by M. de Ferrusac, and has thus taken its place in the scientific world.

Such was fame in America in the year of grace 1820.

In 1817 he did a similar service for systematic entomology, and his contributions to herpetology, to the study of marine invertebrates, especially the crustacea, and to that of invertebrate paleontology, were equally fundamental.

As naturalist of Long's expeditions he described many Western verte-brates, and also collected Indian vocabularies, and it is said that the nar-rative of the expeditions was chiefly based upon the contents of his note books.

In 1825 he removed from Philadelphia to New Harmony, Indiana, and, in company with Maclure and Troost, became a member of the commu-nity founded there by Owen of Lanark. Comparatively little was thence-forth done by him, and we can only regret the untimely close of so brilliant a career.[78]

Charles Alexander Lesueur [b. at Havre-de-Grace, France, January 1, 1778, d. at Havre, December 12, 1846], the friend and associate of Ma-clure and Say, accompanied them to New Harmony. The romantic life of this talented Frenchman has been well narrated in his biography by Ord.[79] He was one of the staff of the Baudin expedition to Australia in 1800, and to his efforts, seconding those of Peron, his associate, were due most of the scientific results which France obtained from that ill-fated enterprise. Lesueur, though a naturalist of considerable ability, was, above all, an artist. The magnificent plates in the reports prepared by Peron[80] and Freycinet[81] were all his. He was called the Raffaelle of zoological painters, and his removal to America in 1815 was greatly deplored by European naturalists. He traveled for three years with Maclure, exploring the West Indies and the eastern United States, making a magnificent collection of drawings of fishes and invertebrates, and in 1818 settled in Philadelphia, where, supporting himself by giving drawing lessons, he became an active member of the Academy of Sciences and published many papers in its Journal.

No one ever drew such exquisite figures of fishes as Lesueur, and it is greatly to be regretted that he never completed his projected work upon North American Ichthyology. He issued a prospectus, with specimen plates, of a Memoir on the Medusae, and his name will always be asso-ciated with the earliest American work upon marine invertebrates and invertebrate paleontology, because it was to him that Say undoubtedly owed his first acquaintance with these departments of zoology. In 1820, while at Albany in the service of the United States and Canadian Boundary Commission, he gave lessons to Eaton and identified his fossils, thus laying the foundations for the future work of the rising school of New York paleontologists.

Twelve years of his life were wasted at New Harmony, and in 1837, after the death of Say, he returned to France, carrying his collections and drawings to the Natural History Museum at Havre, of which he became curator. His period of productiveness was limited to the six years of his residence in Philadelphia. But for their sacrifice to the socialistic ideas of Owen, Say and Lesueur would doubtless be counted among the most distinguished of our naturalists, and the course of American zoological research would have been entirely different.

The Reverend Daniel H. Barnes [b. 1785, d. 1828], of New York, a graduate of Union College and a Baptist preacher, was one of Say's earliest disciples, and from 1823 he published papers on conchology, beginning with an elaborate study of the fresh-water mussels. This group was taken up in 1827 by Doctor Isaac Lea, and discussed from year to year in his well-known series of beautifully illustrated monographs.

Mr. Barnes published, also, papers on the Classification of the Chitonidae, on Batrachian Animals and doubtful Reptiles, and on Magnetic Polarity.

The officers of the Navy had already begun their contributions to natural history which have been so serviceable in later years. One of the earliest contributions by Barnes was a description of five species of *Chiton*, collected in Peru by Captain C. S. Ridgely of the *Constellation*.

In this period (1828 +) was begun the publication of Audubon's folio volumes of illustrations of North American birds—a most extraordinary work, of which Cuvier enthusiastically exclaimed: "C'est le plus magnifique monument que l'Art ait encore élevé a la Nature."

Wilson was the Wordsworth of American naturalists, but Audubon was their Rubens. With pen as well as with brush he delineated those wonderful pictures which have been the delight of the world.

Born in 1781, in Louisiana, while it was still a Spanish colony, he became, at an early age, a pupil of the famous French painter David, under whose tuition he acquired the rudiments of his art. Returning to America, he began the career of an explorer, and for over half a century his life was spent, for the most part, in the forests or in the preparation of his ornithological publications—occasionally visiting England and France, where he had many admirers. His devotion to his work was as complete and self-sacrificing as that of Bowditch, the story of whose translation of La Place has already been referred to. It was a great surprise to his friends (though his own fervor did not permit him to doubt) that the sale of his folio volumes was sufficient to pay his printer's bills. Audubon was not a very accomplished systematic zoologist, and when serious discriminations of species was necessary, sometimes formed alliances with others. Thus Bachman became his collaborator in the study of mammals, and the youthful Baird was invited by him, shortly before his death in 1851, to join him in an ornithological partnership. His relations with Alexander Wilson form the subject of a most entertaining narration in the Ornithological Biography.[82]

Thomas Nuttall [b. in Yorkshire, 1786; d. in St. Helens, Lancashire, September 10, 1859] was so thoroughly identified with American natural history and so entirely unconnected with that of England that, although he returned to his native land to die, we may fairly claim him as one of our own worthies. He crossed the ocean when about twenty-one years of age, and traveled in every part of the United States and in the Sandwich Islands studying birds and plants. From 1822 to 1828 he was curator and lecturer at the Harvard Botanical Garden. Besides numerous papers in the Proceedings of the Philadelphia Academy, he published in Philadelphia, in 1818, his Genera of North American Plants, in his Geological

Sketch of the Valley of the Mississippi, in 1821; his Journal of Travels into the Arkansas Territory, a work abounding in natural history observations; in 1832–1834 his Manual of the Ornithology of the United States and Canada; and in 1843–1849 his North American Sylva, a continuation of the Sylva of Michaux. About 1850 he retired to a rural estate in England, where he died in 1859.

Nuttall was not great as a botanist, as a geologist, or as a zoologist, but was a man useful, beloved, and respected.

Richard Harlan, M. D. [b. 1796, d. 1843], who, with Mitchill, Say, Rafinesque, and Gosse, was one of the earliest of our herpetologists, and who was one of Audubon's chief friends and supporters, published in 1825 the first installment of his Fauna Americana, which treated exclusively of mammals. This was followed, in 1826, by a rival work on mammals by Godman. Harlan's book was a compilation, based largely on translations of portions of Desmarest's Mammalogie, printed three years before in Paris. It was so severely criticised that the second portion, which was to have been devoted to reptiles, was never published, and its author turned his attention to medical literature. Godman's North American National History, or Mastology, contained much original matter, and, though his contemporaries received it with faint praise, it is the only separate, compact, illustrated treatise on the mammals of North America ever published, and is useful to the present day.

John D. Godman [b. in Annapolis, Maryland, December 20, 1794; d. in Germantown, Pennsylvania, April 17, 1830] died an untimely death, but gave promise of a brilliant and useful career as a teacher and investigator. His Rambles of a Naturalist is one of the best series of essays of the Selborne type ever produced by an American, and his American Natural History is a work of much importance, even to the present day, embodying, as it does, a large number of original observations.

Michaux's Sylva was, as we have seen, continued by Nuttall. Wilson's American Ornithology was, in like manner, continued by Charles Lucien

Bonaparte [b. in Paris, May 24, 1803; d. in Paris, July 30, 1857], Prince of Canino, and nephew of Napoleon I, a master in systematic zoology. Bonaparte came to the United States about the year 1822 and returned to Italy in 1828. His contributions to zoological science were of great importance. In 1827 he published in Pisa his Specchio comparativo delle ornithologie di Roma e di Filadelfia, and from 1825 to 1833 his American Ornithology, containing descriptions of over one hundred species of birds discovered by himself.

The publication of Torrey's Flora of the Middle and Northern Sections of the United States was an event of importance, as was also Doctor W. J. Hooker's essay on the Botany of America,[83] the first general treatise upon the American flora or fauna by a master abroad, is pretty sure evidence that the work of home naturalists was beginning to tell.

So also, in a different way, was the appearance in 1829 of the first edition of Mrs. Lincoln's Familiar Lectures on Botany, a work which did much toward swelling the army of amateur botanists.

Important work was also in progress in geology. Eaton and Beck were carrying on the Van Rensselaer survey of New York, and in 1818 the former published his Index to the Geology of the Northern States. Professor Denison Olmsted, of the University of North Carolina, was completing the official survey of that State—the first ever authorized by the government of a State.

Professor Lardner Vanuxem, of North Carolina, in 1828, made an important advance, being the first to avail himself successfully of paleontology for the determination of the age of several of our formations, and their approximate synchronism with European beds.[84]

Horace H. Hayden, of Baltimore [b. 1769, d. 1844], published in 1820 Geological Essays, or an Inquiry into Some of the Geological Phenomena to be Found in Various Parts of America and Elsewhere,[85] which was well received as a contribution to the history of alluvial formations of the globe, and was apparently the first general work on geology published in this country. Silliman said that it should be a text-book in all the schools. He published, also, a New Method of Preserving Anatomical Preparations,[86] A Singular Ore of Cobalt and Manganese,[87] A Description of the Bare Hills near Baltimore,[88] and on Silk Cocoons,[89] and was a founder and vice-president of the Maryland Academy of Sciences.

XV

In the fourth decade (1830–1840) the leading spirits were Silliman, Hare, Olmsted, Hitchcock, Torrey, De Kay, Henry, and Morse.

Among the men just coming into prominence were J. W. Draper, then professor in Hampden Sidney College, in Virginia, the brothers W. B. and H. D. Rogers, A. A. Gould, the conchologist, and James D. Dana.

Henry was just making his first discoveries in physics, having in 1829 pointed out the possibility of electro-magnetism as a motive power, and in 1831 set up his first telegraphic circuit at Albany. In 1832 the United Coast Survey, discontinued in 1818, was reorganized under the direction of its first chief, Hassler, now advanced in years.[90]

The natural-history survey of New York was organized by the State in 1836, and James Hall and Ebenezer Emmons were placed upon its staff.

G. W. Featherstonhaugh [b. 1780, d. 1866] was conducting (1834–35) a Government expedition, exploring the geology of the elevated country between the Missouri and Red rivers, and the Wisconsin territories. He bore the name of United States Geologist, and projected a geological map of the United States, which now, half a century later, is being completed by the United States geologist of to-day. Beside his report upon the survey just referred to, Featherstonhaugh printed a Geological Reconnaissance, in 1835, from Green Bay to Coteau des Prairies, and a Canoe Voyage up the Minnay Sotor, in London, 1847.

In 1838 the United States Exploring Expedition under Wilkes was sent upon its voyage of circumnavigation, having upon its staff a young naturalist named Dana, whose studies upon the crustaceans and radiates of the expedition have made him a world-wide reputation, entirely independent of that which he has since gained as a mineralogist and geologist.

It is customary to refer to the Wilkes expedition as having been sent out entirely in the interests of science. As a matter of fact it was organized primarily in the interests of the whale fishery of the United States.

Dana, before his departure with Wilkes, had published, in 1837, the first edition of his System of Mineralogy, a work which, in its subsequent editions, has become the standard manual of the world.

The publication of Lyell's Principles of Geology at the beginning of this decade (1830) had given new direction to the thoughts of our geologists, and they were all hard at work under its inspiration.

With 1839 ended the second of our thirty-year periods—the one which I have chosen to speak of as the period of Silliman—not so much because of the investigations of the New Haven professor, as on account of his influence in the promotion of American science and scientific institutions.

This was a time of hard work, and we must not withhold our praise from the noble little company of pioneers who were in those years building the foundations upon which the scientific institutions of to-day are resting.

The difficulties and drawbacks of scientific research at this time have been well described by one who knew them:[91]

The professedly scientific institutions of our country issued from time to time, though at considerable intervals, volumes of Transactions and Proceedings unquestionably not without their influence in keeping alive the scarcely kindled flame, but whose contents, as might be expected, were for the most part rather in conformity with the then existing standard of excellence than in advance of it. Natural History in the United States was the mere sorting of genera and species; the highest requisite for distinction in any physical science was the knowledge of what European students had attained;—astronomy was in general confined to observations, and those not of the most refined character, and its merely descriptive departments were estimated far more highly than the study of its laws. Astronomical computation had hardly risen above the ciphering out of eclipses and occultations. Indeed, I risk nothing in saying that Astronomy had lost ground in America since those Colonial times when men like Rittenhouse kept up a constant scientific communication with students of Astronomy beyond the seas. And I believe I may further say, that a single instance of a man's devoting himself to science as the only earthly guide, aim, and object of his life, while unassured of a professor's chair or some analogous appointment, upon which he might depend for subsistence, was utterly unknown.

Such was the state of science in general. In Astronomy the expensive appliances requisite for all observations of the higher class were wanting, and there was not in the United States, with the exception of the Hudson Observatory, to which Professor Loomis devoted such hours as he could spare from his duties in the College, a single establishment provided with the means of making an absolute determination of the place of any celestial body, or even relative determinations at all commensurate in accuracy with the demands of the times. The only instrument that could be thought of for the purpose was the Yale-College telescope, which, although provided with a micrometer, was destitute of the means of identifying comparison—stars. A better idea of American astronomy a dozen years ago can hardly be obtained than by quoting from an article published at that time by the eminent geometer who now retires from the position of President of this Association. He will forgive me the liberty, for the sake of the illustration. "The impossibility," said he, "of great national progress in Astronomy while the materials are for the most part imported can hardly need to be impressed upon the patrons of science in this country. . . . And next to the support of observers is the establishment of observatories. Something has been done for this purpose in various parts of the country, and it is earnestly to be hoped that the intimations which we have

heard regarding the intentions of government may prove to be well
founded;—that we shall soon have a permanent national observatory equal in
its appointments to the best furnished ones of Europe; and that American
ships will ere long calculate their longitudes and latitudes from an American
nautical almanac. That there is on this side of the Atlantic a sufficient capacity
for celestial observations is amply attested by the success which has attended
the efforts, necessarily humble, which have hitherto been made."[92]

XVI

Just before the middle of the century a wave, or, to speak more accurately, a series of waves, of intellectual activity began to pass over Europe
and America. There was a renaissance quite as important as that which
occurred in Europe at the close of the middle ages. Draper and other
historians have pointed out the causes of this movement, prominent
among which were the introduction of steam and electricity, annihilating
space and relieving mankind from a great burden of mechanical drudgery. It was the beginning of the "age of science," and political as well as
social and industrial changes followed in rapid succession.

　　In Europe the great work began a little earlier. Professor Huxley, in
his address to the Royal Society in 1885, took for a fixed point his own
birthday in 1825, which was four months before the completion of the
railway between Stockton and Darlington—"the ancestral representative
of the vast reticulated fetching and carrying organism which now extends
its meshes over the civilized world."

Since then, [he remarked,] the greater part of the vast body of knowledge
which constitutes the modern sciences of physics, chemistry, biology, and geology has been acquired, and the widest generalizations therefrom have been
deduced, and, furthermore, the majority of those applications of scientific

knowledge to practical ends which have brought about the most striking differences between our present civilization and that of antiquity have been made within that period of time.

It is within the past half century, he continued, that the most brilliant additions have been made to fact and theory and serviceable hypothesis in the region of pure science, for within this time falls the establishment on a safe basis of the greatest of all the generalizations of science, the doctrines of the Conservation of Energy and of Evolution. Within this time the larger moiety of our knowledge of light, heat, electricity, and magnetism has been acquired. Our present chemistry has been, in great part, created, while the whole science has been remodeled from foundation to roof.

It may be natural [continued Professor Huxley] that progress should appear most striking to me among those sciences to which my own attention has been directed, but I do not think this will wholly account for the apparent advance "by leaps and bounds" of the biological sciences within my recollection. The cell theory was the latest novelty when I began to work with the microscope, and I have watched the building of the whole vast fabric of histology. I can say almost as much of embryology, since Von Baer's great work was published in 1828. Our knowledge of the morphology of the lower plants and animals and a great deal of that of the higher forms has very largely been obtained in my time; while physiology has been put upon a totally new foundation and, as it were, reconstructed, by the thorough application of the experimental method to the study of the phenomena of life, and by the accurate determination of the purely physical and chemical components of these phenomena. The exact nature of the processes of sexual and nonsexual reproduction has been brought to light. Our knowledge of geographical and geological distribution and of the extinct forms of life has been increased a hundredfold. As for the progress of geological science, what more need be said than that the first volume of Lyell's Principles bears the date of 1830.

It can not be expected that, within the limits of this address, I should attempt to show what America has done in the last half century. I am striving to trace the beginnings, not the results, of scientific work on this side of the Atlantic. I will simply quote what was said by the London Times in 1876:

In the natural distribution of subjects the history of enterprise, discovery, and conquest, and the growth of republics fell to America, and she has dealt nobly with them. In the wider and more multifarious provinces of art and science she runs neck and neck with the mother country and is never left behind.

It is difficult to determine exactly the year when the first waves of this renaissance reached the shores of America. Silliman, in his Priestley address, placed the date at 1845. I should rather say 1840, when the first national scientific association was organized, although signs of awakening may be detected even before the beginning of the previous decade. We must, however, carefully avoid giving too much prominence to the influence of individuals. I have spoken of this period of thirty years as the period of Agassiz. Agassiz, however, did not bring the waves with him; he came in on the crest of one of them; he was not the founder of modern American natural history, but as a public teacher and organizer of institutions, he exerted a most important influence upon its growth.

One of the leading events of the decade was the reorganization of the Coast Survey in 1844, under the sage administration of Alexander Dallas Bache,[93] speedily followed by the beginning of investigations upon the Gulf Stream, and of the researches of Count Pourtalès into its fauna, which laid the foundations of modern deep-sea exploration. Others were the founding of the Lawrence Scientific School, the Cincinnati Observatory, the Yale Analytical Laboratory, the celebration of the Centennial Jubilee of the American Philosophical Society in 1843, and the enlargement of Silliman's American Journal of Science.

The Naval Astronomical Expedition was sent to Chili, under Gilliss (1849), to make observations upon the parallax of the sun. Lieutenant Lynch was sent to Palestine (in 1848) at the head of an expedition to explore the Jordan and the Dead Sea.

Frémont conducted expeditions, in 1848, to explore the Rocky Mountains and the territory beyond; and Stansbury, in 1849–50, a similar exploration of the valley of the Great Salt Lake. David Dale Owen was heading a Government geological survey in Wisconsin, Iowa, and Minnesota (1848), and from all of these came results of importance to science and to natural history.

In 1849 Professor W. H. Harvey, of Dublin, visited America and collected materials for his Nereis Boreali-Americana, which was the foundation of our marine botany.

Sir Charles Lyell, ex-president of the Geological Society of London, visited the United States in 1841 and again in 1845, and published two volumes of travels, which were, however, of much less importance than the effects of his encouraging presence upon the rising school of American geologists. His Principles of Geology, as has already been said, was an epoch-making work, and he was to his generation almost what Darwin was to the one which followed.

Certain successes of our astronomers and physicists had a bearing upon the progress of American science in all its departments which was, perhaps, even greater than their actual importance would seem to warrant. These were the discovery, by the Bards of Cambridge, of Bards comet in 1846, of the satellite Hyperion in 1848, of the third ring of Saturn in 1850, the discovery by Herrick and Bradley, in 1846, of the bipartition of Belas comet, and the application of the telegraph to longitude determination after Locke had constructed (in 1848) his clock for the registration of time observations by means of electro-magnetism.

It is almost ludicrous at this day to observe the grateful sentiments with which our men of science welcomed the adoption of this American method in the observatory at Greenwich. Americans were still writhing under the sting of Sidney Smith's demand, Who reads an American book? and the narrations of those critical observers of national customs, Dickens, Basil Hall, and Mrs. Trollope. The continental approval of American science was like balsam to the sensitive spirits or our countrymen.

John William Draper's versatile and original researches in physics were also yielding weighty results, and as early as 1847 he had already laid the foundations of the science of spectroscopy, which Kirchhoff so boldly appropriated many years later.

Most important of all, by reason of its breadth of scope, was the foundation of the Smithsonian Institution, which was organized in 1846 by the election of Joseph Henry to its secretaryship. Who can attempt to say what the conditions of science in the United States would be to-day but for the bequest of Smithson? In the words of John Quincy Adams "Of all the foundations or establishments for pious or charitable uses which ever signalized the spirit of the age or the comprehensive beneficence of the founder, none can be named more deserving the approbation of mankind."

Among the leaders of this new enterprise and of the scientific activities of the day may be named Silliman, Hare, Henry, Bache, Maury, Alexander, Locke, Mitchel, Peirce, Walker, Draper, Dana, Wyman, Agassiz, Gray, Torrey, Haldeman, Morton, Holbrook, Gibbes, Gould, DeKay, Storer, Hitchcock, Redfield, the brothers Rogers, Jackson, Hays, and Owen.

Among the rising men were Baird, Adams the conchologist, Burnett, Harris the entomologist, and the LeConte brothers among zoologists; Lapham, D. C. Eaton, and Grant, among botanists; Sterry Hunt, Brush, J. D. Whitney, Wolcott Gibbs, and Lesley, among chemists and geologists, as well as Schiel, of St. Louis, who had before 1842 discovered the principle of chemical homology.

I have not time to say what ought to be said of the coming of Agassiz in 1846. He lives in the hearts of his adopted countrymen. He has a colossal monument in the museum which he reared, and a still greater one in the lives and works of pupils such as Agassiz, Allen, Burgess, Burnett, Brooks, Clarke, Cooke, Faxon, Fewkes, Gorman, Hartt, Hyatt, Joseph LeConte, Lyman, McCrady, Morse, Mills, Niles, Packard, Putnam, Scudder, St. John, Shaler, Verrill, Wilder, and David A. Wells.

XVII

They were glorious men who represented American science at the middle of the century. We may well wonder whether the present decade will make as good a showing forty years hence.

The next decade was its continuation. The old leaders were nearly all active, and to their ranks were added many more.

An army of new men was rising up.

It was a period of great explorations, for the frontier of the United States was sweeping westward, and there was need of a better knowledge of the public domain.

Sitgreaves explored the region of the Zuni and Colorado rivers in 1852, and Marcy the Red River of the North. The Mexican boundary survey, under Emory, was in progress from 1854 to 1856, and at the same time the various Pacific railroad surveys. There was also the Herndon exploration of the valley of the Amazon, and the North Pacific exploring expedition under Rogers. These were the days, too, when that extensive exploration of British North America was begun through the cooperation of the Hudson's Bay Company with the Smithsonian Institution.

It was the harvest time of the museums. Agassiz was building up with immense rapidity his collections in Cambridge, utilizing to the fullest extent the methods which he had learned in the great European establishments and the public spirit and generosity of the Americans. Baird was using his matchless powers of organization in equipping and inspiring the officers of the various surveys and accumulating immense collections to be used in the interest of the future National Museum.

Systematic natural history advanced with rapid strides. The magnificent folio reports of the Wilkes expedition were now being published, and some of them, particularly those by Dana on the crustaceans and the zoophytes and geology, that of Gould upon the mollusks, those by Torrey, Gray, and Eaton upon the plants, were of great importance.

The reports of the domestic surveys contained numerous papers upon systematic natural history, prepared under the direction of Baird, assisted by Girard, Gill, Cassin, Suckley, LeConte, Cooper, and others. The volumes relating to the mammals and the birds, prepared by Baird's own pen, were the first exhaustive treatises upon the mammalogy and ornithology of the United States.

The American Association was doing a great work in popular education through its system of meeting each year in a different city. In 1850 it met in Charleston, and its entire expenses were paid by the city corporation as a valid mark of public approval, while the foundation of the Charleston Museum of Natural History was one of the direct results of the meeting.

In 1857 it met in Montreal, and delegates from the English scientific societies were present. This was one of the earliest of those manifestations of international courtesy upon scientific ground of which there have since been many.

In the seventh decade, which began with threatenings of civil war, the growth of science was almost arrested. A meeting of the American Association was to have been held in Nashville in 1861, but none was called. In 1866, at Buffalo, its sessions were resumed with the old board of officers elected in 1860. One of the vice-presidents, Gibbes, of South Carolina, had not been heard from since the war began, and the Southern members were all absent. Many of the Northern members wrote explaining that they could not attend this meeting because they could not afford it, "such has been the increase of living expenses, without a corresponding increase in the salaries of men of science." Few scientists were engaged in the war, though one, O. M. Mitchel, who left the directorship of the Dudley Observatory to accept the command of an Ohio brigade, died in service in 1862, and another, Couthouy, sacrificed his life in the Navy. Others, like Ordway, left the ranks of science never to resume their places as investigators.

Scientific effort was paralyzed, and attention was directed to other matters. In 1864, when the Smithsonian building was burned, Lincoln, it is said, looking at the flames from the windows of the Executive Mansion, remarked to some military officers who were present: "Gentlemen, yonder is a national calamity. We have no time to think about it now; we must attend to other things."

The only important events during the war were two; one the organization of the National Academy of Sciences, which soon became what Bache had remarked the necessity for in 1851, when he said: An institution of science, supplementary to existing ones, is much needed in our country to guide public action in reference to scientific matters.[94]

The other was the passage, in 1862, of the bill for the establishment of scientific educational institutions in every State. The agricultural colleges were then, as they still are, unpopular among many scientific men, but the wisdom of the measure is apparently before long to be justified.

Before the end of the decade the Northern States[95] had begun a career of renewed prosperity, and the scientific institutions were reorganized. The leading spirits were such men as Pierce, Henry, Agassiz, Gray, Barnard, the Goulds, Newberry, Lea, Whittlesey, Foster, Rood, Cooke, Newcomb, Newton, Wyman, Winchell.

Among the rising men, some of them very prominent before 1870, were Barker, Bolton, Chandler, Egleston, Hall, Harkness, Langley, Mayer, Pickering, Young, Powell, Pumpelly, Abbe, Collett, Emerson, Hartt, Lupton, Marsh, Whitfield, Williams, N. H. Winchell, Agassiz, the Allens, Beale, Cope, Cones, Canby, Dall, Hoy, Hyatt, Morse, Orton, Perkins, Rey, Riley, Scudder, Sidney Smith, Sterns, Tuttle, Verrill, Wood.

Soon after the war the surveys of the West, which have coalesced to form the United States Geological Survey, were forming under the direction of Clarence King, Lieutenant Wheeler, F. V. Hayden, and Major Powell.

The discovery of the nature of the corona of the sun by Young and Harkness in 1869 was an event encouraging to the rising spirits of our workers.

XVIII

With 1869 we reach the end of the third period and the threshold of that in which we are living. I shall not attempt to define the characteristics of the natural history of to-day, though I wish to direct attention to certain tendencies and conditions which exist. Let me, however, refer once more to the past, since it leads again directly up to the present.

In a retrospect published in 1876,[96] one of our leaders stated that American science during the first forty years of the present century was in "a state of general lethargy, broken now and then by the activity of some first-class man, which, however, commonly ceased to be directed into purely scientific channels." This depiction was, no doubt, somewhat true of the physical and mathematical sciences concerned, but not to the extent indicated by the writer quoted. What could be more unjust to the men of the last generation than this? "It is," continues he, "strikingly illustrative of the absence of everything like an effective national pride in science that two generations should have passed without America having produced anyone to continue the philosophical researches of Franklin."

I may not presume to criticise the opinion of the writer from whom these words are quoted, but I can not resist the temptation to repeat a paragraph from Professor John W. Draper's eloquent centennial address upon Science in America:

In many of the addresses on the centennial occasion [he said], the shortcom-
ings of the United States in extending the boundaries of scientific knowledge,
especially in the physical and chemical departments, have been set forth. "We
must acknowledge with shame our inferiority to other people," says one. "We
have done nothing," says another. . . . But we must not forget that many of
these humiliating accusations are made by persons who are not of authority in
the matter; who, because they are ignorant of what has been done, think that
nothing has been done. They mistake what is merely a blank in their own
information for a blank in reality. In their alacrity to depreciate the merit of
their own country they would have us confess that, for the last century, we
have been living on the reputation of Franklin and his thunder rod.

These are the words of one who, himself an Englishman by birth,
could, with excellent grace, upbraid our countrymen for their lack of
patriotism.

The early American naturalists have been reproached for devoting
their time to explorations and descriptive natural history, and their work
depreciated, as being of a character beneath the dignity of the biologists
of to-day.

The zoological science of the country, [said the president of the Natural History section of the American Association a few years since], presents itself in
two distinct periods: The first period may be recognized as embracing the lowest stages of the science; it included, among others, a class of men who busied
themselves in taking an inventory of the animals of the country, an important
and necessary work to be compared to that of the hewers and diggers who
first settle a new country, but in their work demanded no deep knowledge or
breadth of view.

It is quite unnecessary to defend systematic zoology from such slurs
as this, nor do I believe that the writer quoted would really defend the
ideas which his words seem to convey, although, as Professor Judd has

regretfully confessed in his recent address before the Zoological Society of London, systematic zoologists and botanists have become somewhat rare and out of fashion in Europe in modern times.

The best vindication of the wisdom of our early writers will be, I think, the presentation of a counter quotation from another presidential address, that of the venerable Doctor Bentham before the Linnæan Society of London, in 1867:

It is scarcely half a century [wrote Bentham], since our American brethren applied themselves in earnest to the investigation of the natural productions and physical condition of their vast continent, their progress, especially during the latter half of that period, had been very rapid until the outbreak of the recent war, so deplorable in its effects in the interests of science as well as on the material prosperity of their country. The peculiar condition of the North American Continent requires imperatively that its physical and biological statistics should be accurately collected and authentically recorded, and that this should be speedily done. It is more than any country, except our Australian colonies, in a state of transition. Vast tracts of land are still in what may be called almost a primitive state unmodified by the effects of civilization, uninhabited, or tenanted only by the remnants of ancient tribes, whose unsettled life never exercised much influence over the natural productions of the country. But this state of things is rapidly passing away; the invasion and steady progress of a civilized population, whilst changing generally the face of nature, is obliterating many of the evidences of a former state of things. It may be true that the call for recording the traces of previous conditions may be particularly strong in Ethnology and Archæology; but in our own branches of the science, the observations and consequent theories of Darwin having called special attention to the history of species, it becomes particularly important that accurate biological statistics should be obtained for future comparison in those countries where the circumstances influencing those conditions are the most rapidly changing. The larger races of wild animals are dwindling down, like the aboriginal inhabitants, under the deadly influence of civilized man. Myriads of the lower orders of animal life, as well as of plants, disappear with the destruction of forests, the drainage of swamps, and the gradual spread of cultivation, and their places are occupied by foreign invaders. Other races, no doubt, without actually disappearing, undergo a gradual change under the new order of things, which, if perceptible only in the course of successive generations, require so much the more for future proof an accurate record of their state in the still unsettled conditions of the country. In the Old World almost every attempt to compare the present state of vegetation or animal life with that which existed in uncivilized times is in a great measure frustrated by the absolute want of evidence as to that former state; but in North America the change is going forward as it were close under the eye of the observer. This consideration may one day give great value to the reports of the naturalist sent by the Government, as we have seen, at the instigation of the Smithsonian Institution and other promoters of science, to accompany the surveys of new territories.

Having said this much in defense of the scientific men of the United States, I wish, in conclusion, to prefer some very serious charges against the country at large, or, rather, as a citizen of the United States, to make some very melancholy and humiliating confessions.

The present century is often spoken of as "the age of science," and Americans are somewhat disposed to be proud of the manner in which scientific institutions are fostered and scientific investigators encouraged on this side of the Atlantic.

Our countrymen have made very important advances in many departments of research. We have a few admirably organized laboratories and observatories, a few good collections of scientific books, six or eight museums worthy of the name, and a score or more of scientific and technological schools, well organized and better provided with officers than with money. We have several strong scientific societies, no one of which, however, publishes transactions worthy of its own standing and the collective reputation of its members. In fact, the combined publishing funds of all our societies would not pay for the annual issue of a volume of memoirs such as appears under the auspices of any one of a dozen European societies which might be named.

Our Government, by a liberal support of its scientific departments, has done much to atone for the really feeble manner in which local institutions have been maintained. The Coast Survey, the Geological Survey, the Department of Agriculture, the Fish Commission, the Army, with its Meteorological Bureau, its Medical Museum and Library, and its explorations; the Navy, with its Observatory, its laboratories and its explorations; and, in addition to these, the Smithsonian Institution, with its systematic promotion of all good works in science, have accomplished more than is ordinarily placed to their credit. Many hundreds of volumes of scientific memoirs have been issued from the Government Printing Office since 1870, and these have been distributed in such a generous and far-reaching way that they have not failed to reach every town and village in the United States where a roof has been provided to protect them.

It may be that some one will accuse the Government of having usurped the work of the private publisher. Very little of value in the way of scientific literature has been issued during the same period by publishers, except in reprints or translations of works of foreign investigators. It should be borne in mind, however, that our Government has not only published the results of investigations, but has supported the investigators and provided them with laboratories, instruments, and material, and that the memoirs which it has issued would never, as a rule, have been accepted by private publishers.

I do not wish to underrate the efficiency of American men of science, nor the enthusiasm with which many public men and capitalists have promoted our scientific institutions. Our countrymen have had wonderful successes in many directions. They have borne their share in the battle of science against the unknown. They have had abundant recognition from their fellow-workers in the Old World. They have met perhaps a more intelligent appreciation abroad than at home. It is the absence of home appreciation that causes us very much foreboding in the future.

In Boston or Cambridge, in New York, Philadelphia, Baltimore, Washington, Chicago, or San Francisco, and in most of the college towns, a man interested in science may find others ready to talk over with him a new scientific book or a discovery which has excited his interest. Elsewhere the chances are he will have to keep his thoughts to himself. One may quickly recite the names of the towns and cities in which may be found ten or more people whose knowledge of any science is aught than vague and rudimentary. Let me illustrate my idea by supposing that every inhabitant of the United States over fifteen years of age should be required to mention ten living men eminent in scientific work, would one out of a hundred be able to respond? Does anyone suppose that there are three or four hundred thousand people enlightened to this degree?

Let us look at some statistics, or rather some facts, which it is convenient to arrange in statistical form. The total number of white inhabitants of the United States in 1880 was about 42,000,000. The total number of naturalists, as shown in the Naturalist's Directory for 1886, was a little over 4,600. This list includes not only the investigators, who probably do not exceed 500 in number, and the advanced teachers, who muster perhaps 1,000 strong, but all who are sufficiently interested in science to have selected special lines of study.

We have, then, 1 person interested in science to about 10,000 inhabitants. But the leaven of science is not evenly distributed through the national loaf. It is the tendency of scientific men to congregate together. In Washington, for instance, there is 1 scientific man to every 500 inhabitants; in Cambridge, 1 to 830; and in New Haven, 1 to 1,100. In New Orleans the proportion is 1 to 8,800; in Jersey City, 1 to 24,000; in New York, 1 to 7,000; and in Brooklyn, 1 to 8,500. I have before me the proportions worked out for the seventy-five principal cities of the United States. The showing is suggestive, though no doubt in some instances misleading. The tendency to gregariousness on the part of scientific men may perhaps be further illustrated by a reference to certain societies. The membership of the National Academy of Sciences is almost entirely concentrated about Boston, New York, Philadelphia, Washington, and New Haven. Missouri has one member, Illinois one, Ohio one, Maryland, New Jersey, and Rhode Island three, and California four, while thirty-two States and Territories are not represented. A precisely similar distribution of members is found in the American Society of Naturalists. A majority of the members of the American Association for the Advancement of Science live in New York, Massachusetts, Pennsylvania, the District of Columbia, Michigan, Minnesota, Ohio, Illinois, and New Jersey.

It has been stated that the average proportion of scientific men to the population at large is 1 to 10,000. A more minute examination shows that while fifteen of the States and Territories have more than the average proportion of scientific men, thirty-two have less. Oregon and California, Michigan and Delaware have very nearly the normal number. Massachusetts, Rhode Island, Connecticut, Illinois, Colorado, and Florida have about 1 to 4,000. West Virginia, Nevada, Arkansas, Mississippi, Georgia, Kentucky, Texas, Alabama, and the Carolinas are the ones least liberally furnished. Certain cities appear to be absolutely without scientific men. The worst cases of destitution seem to be Paterson, New Jersey, a city of 50,000 inhabitants; Wheeling, West Virginia, with 30,000; Quincy, Illinois, with 26,000; Newport, Kentucky, with 20,000; Williamsport, Pennsylvania, and Kingston, New York, with 18,000; Council Bluffs, Iowa, and Zanesville, Ohio, with 17,000; Oshkosh, Wisconsin, and Sandusky, Ohio, with 15,000; Lincoln, Rhode Island; Norwalk, Connecticut, and Brockton and Pittsfield, Massachusetts, with 13,000. In these there are no men of science recorded, and eight cities of more than 15,000 inhabitants have only one, namely, Omaha, Nebraska, and St. Joseph, Missouri; Chelsea, Massachusetts; Cohoes, New York; Sacramento, California; Binghamton, New York; Portland, Oregon; and Leadville, Colorado.

Of course these statistical statements are not properly statistics. I have no doubt that some of these cities are misrepresented in what has been said. This much, however, is probably true, that not one of them has a scientific society, a museum, a school of science, or a sufficient number of scientific men to insure even the occasional delivery of a course of scientific lectures.

Studying the distribution of scientific societies, we find that there are fourteen States and Territories in which there are no scientific societies whatever. There are fourteen States which have State academies of science or societies which are so organized as to be equivalent to State academies.

Perhaps the most discouraging feature of all is the diminutive circulation of scientific periodicals. In addition to a certain number of specialists' journals, we have in the United States three which are wide enough in scope to be necessary to all who attempt to keep an abstract of the progress of science. Of these the American Journal of Science has, we are told, a circulation of less than 800; the American Naturalist less than 1,100, and Science less than 6,000. A considerable proportion of the copies printed go, as a matter of course, to public institutions, and not to individuals. Even the Popular Science Monthly and the Scientific American, which appeal to large classes of unscientific readers, have circulations absurdly small.

The most effective agents for the dissemination of scientific intelligence are probably the religious journals, aided to some extent by the agricultural journals and to a very limited degree by the weekly and daily newspapers. It is much to be regretted that several influential journals, which ten or fifteen years ago gave attention to the publication of trustworthy scientific intelligence, have of late almost entirely abandoned the effort. The allusions to science in the majority of our newspapers are singularly inaccurate and unscholarly, and too often science is referred to only when some of its achievements offer opportunity for witticism.

The statements which I have just made may, as I have said, prove in some instances erroneous and to some extent misleading, but I think the general tendency of a careful study of the distribution of scientific men and institutions is to show that the people of the United States, except in so far as they sanction by their approval the work of the scientific departments of the Government and the institutions established by private munificence, have little reason to be proud of the national attitude toward science.

I am, however, by no means despondent for the future. The importance of scientific work is thoroughly appreciated, and it is well understand that many important public duties can be performed properly only by trained men of science. The claims of science to a prominent place in every educational plan are every year more fully conceded. Science is permeating the theory and the practice of every art and every industry, as well as every department of learning. The greatest danger to science is perhaps the fact that all who have studied at all within the last quarter of a century have studied its rudiments and feel competent to employ its methods and its language and to form judgments on the merits of current work.

In the meantime the professional men of science, the scholars, and the investigators seem to me to be strangely indifferent to the questions as to how the public at large is to be made familiar with the results of their labors. It may be that the tendency to specialization is destined to deprive the sciences of their former hold upon popular interest, and that the study of zoology, botany and geology, mineralogy and chemistry, will become so technical, that each will require the exclusive attention of its votaries for a period of years. It may be that we are to have no more zoologists such as Agassiz and Baird, no more botanists such as Gray, and that the place which such men filled in the community will be supplied by combinations of a number of specialists, each of whom knows, with more minuteness, limited portions of the subjects grasped bodily by the masters of the last generation. It may be that the use of the word naturalist is to become an anachronism, and that we are all destined to become generically biologists, and specifically morphologists, histologists, embryologists, physiologists, or it may be cetologists, chiropterologists, oologists, carcinologists, ophiologists, helminthologists, actinologists, coleopterists, caricologists, mycologists, muscologists, bacteriologists, diatomologists, paleobotanists, crystallographers, petrologists, and the like.

I can but believe, however, that it is the duty of every scientific scholar, however minute his specialty, to resist in himself and in the professional circles which surround him, the tendency toward narrowing technicality in thought and sympathy, and above all in the education of nonprofessional students.

I can not resist the feeling that American men of science are in a large degree responsible if their fellow-citizens are not fully awake to the claims of scientific endeavor in their midst.

I am not in sympathy with those who feel that their dignity is lowered when their investigations lead toward improvement in the physical condition of mankind, but I feel that the highest function of science is to minister to their mental and moral welfare. Here in the United States, more than in any other country, it is necessary that sound, accurate knowledge and a scientific manner of thought should exist among the people, and the man of science is becoming more than ever the natural custodian of the treasured knowledge of the world. To him above all others falls the duty of organizing and maintaining the institutions for the diffusion of knowledge, many of which have been spoken of in these addresses—the schools, the museums, the expositions, the societies, the periodicals. To him more than to any other American should be made familiar the words of President Washington in his farewell address to the American people:

"Promote, then, as an object of primary importance, institutions for the general diffusion of knowledge. In proportion as the structure of a government gives force to public opinions it should be enlightened."

THE ORIGIN OF THE NATIONAL SCIENTIFIC AND EDUCATIONAL INSTITUTIONS OF THE UNITED STATES[1]

By George Brown Goode

Assistant Secretary, Smithsonian Institution, in charge of the U.S. National Museum

"Early in the seventeenth century," we are told, "the great Mr. Boyle, Bishop Wilkins, and several other learned men proposed to leave England and establish a society for promoting knowledge in the new colony [of Connecticut], of which Mr. Winthrop,[2] their intimate friend and associate, was appointed governor."

"Such men," wrote the historian, "were too valuable to lose from Great Britain, and Charles the Second having taken them under his protection in 1661, the society was there established, and received the title of The Royal Society of London."[3]

For more than a hundred years this society was for our country what it still is for the British colonies throughout the world—a central and national scientific organization. All Americans eminent in science were on its list of Fellows, among them Cotton Mather, the three Winthrops, Bowdoin, and Paul Dudley, in New England; Franklin, Rittenhouse, and Morgan, in Pennsylvania; Banister, Clayton, Mitchell, and Byrd, in Virginia; and Garden and Williamson in the Carolinas, while in its Philosophical Transactions were published the only records of American research.[4]

It was not until long after the middle of the last century that any scientific society was permanently established in North America, although serious but fruitless efforts were made in this direction as early as 1743, when Benjamin Franklin issued his circular entitled A proposal for promoting useful knowledge among the British plantations in America, in which it was urged "that a society should be formed of *virtuosi* or ingenious men residing in the several colonies, to be called *The American Philosophical Society*."

There is still in existence, in the possession of the Philosophical Society in Philadelphia, a most interesting letter from Franklin to Governor Cadwallader Colden, of New York, in which he tells of the steps which had already been taken for the formation of a scientific society in Philadelphia, and of the means by which he hoped to make it of great importance to the colonies.

Our forefathers were not yet prepared for the society, nor for the American Philosophical Miscellany which Franklin proposed to issue, either monthly or quarterly. There is no reason to believe that the society ever did anything of importance. Franklin's own attention was soon directed exclusively to his electrical researches, and his society languished and died.

Some twenty years later, in 1766, a new organization was attempted under the title of The American Society held at Philadelphia for Promoting Useful Knowledge.[5] Franklin, although absent in England, was elected its president, and the association entered upon a very promising career.

In the meantime the few surviving members of the first American Philosophical Society formed, under the old name, an organization which in many particulars was so unlike that proposed in 1743 that it might almost be regarded as new rather than a revival. Its membership included many of the most influential and wealthy colonists, and the spirited manner in which it organized a plan for the observation of the transit of Venus in 1769 gave it at once a respectable standing at home and abroad.

In 1769, after negotiations which occupied nearly a year, the two societies were united,[6] and The American Philosophical Society held at Philadelphia for Promoting Useful Knowledge has from that time until now maintained an honorable position among the scientific organizations of the world.

The society at once began the publication of a volume of memoirs, which appeared in 1771 under the name of The American Philosophical Transactions.[7]

From 1773 to 1779 its operations were often interrupted. In the minutes of the meeting for December, 1774, appears the following remarkable note in the handwriting of Doctor Benjamin Rush, one of the secretaries, soon after to be one of the signers of the Declaration of Independence:

The act of the British Parliament for shutting up the port of Boston, for altering the charters, and for the more impartial administration of justice in the Province of Massachusetts Bay, together with a bill for establishing Popery and arbitrary power in Quebec, having alarmed the whole of the American colony, the members of the American Philosophical Society, partaking with their countrymen in the distress and labors brought upon their country, were obliged to discontinue their meetings for some months until a mode of opposition to the said acts of Parliament was established, which we hope may restore the former harmony and maintain a perpetual union between Great Britain and the Americas.

This entry is especially interesting because it emphasizes the fact that among the members of this infant scientific society were many of the men who were most active in the organization of the Republic, and who, under the stress of the times, abandoned the quiet pursuits of science and devoted themselves to the national interests which were just coming into being.

Franklin was president from its organization until his death, in 1790. He was at the same time president of the Commonwealth of Pennsylvania and a member of the Constitutional Convention, and the eminence of its leader probably secured for the body greater prestige than would otherwise have been attainable. The society, in fact, soon assumed national importance, for, during the last decade of the century and for many years after, Philadelphia was the metropolis of American science and literature.

Directly after the Revolution a similar institution was established in Boston, the American Academy of Arts and Sciences, which was incorporated by the legislature of Massachusetts in 1780, and published its first memoirs in 1785. This, like the Philadelphia society, owed its origin to the efforts of a great statesman. We find the whole history in the memoirs of John Adams, a man who believed, with Washington, that scientific institutions are the best and most lasting protection of a popular government.

In a memorandum written in 1809 Mr. Adams gave his recollections of the circumstances which led to his deep and lasting interest in scientific foundations:

In traveling from Boston to Philadelphia, in 1774–75–76–77, I had several times amused myself at Norwalk, in Connecticut, with the very curious collection of birds and insects of American production made by Mr. Arnold;[8] a collection which he afterwards sold to Governor Tryon, who sold it to Sir Ashton Lever, in whose apartments in London I afterwards viewed it again. This collection was so singular a thing that it made a deep impression upon me, and I could not but consider it a reproach to my country that so little was known, even to herself, of her natural history.

When I was in Europe, in the years 1778–79 in the commission to the King of France, with Doctor Franklin and Mr. Arthur Lee, I had opportunities to see the King's collection and many others, which increased my wishes that nature might be examined and studied in my own country as it was in others.

In France, among the academicians and other men of science and letters, I was frequently entertained with inquiries concerning the Philosophical Society of Philadelphia, and with eulogiums on the wisdom of that institution, and encomiums on some publications in their transactions. These conversations suggested to me the idea of such an establishment in Boston, where I knew there was as much love of science, and as many gentlemen who were capable of pursuing it, as in any other city of its size.

In 1779 I returned to Boston on the French frigate *La Sensible*, with the Chevalier de la Luzerne and M. Marbois.[9] The corporation of Harvard College gave a public dinner in honor of the French ambassador and his suite, and did me the honor of an invitation to dine with them. At table in the Philosophy Chamber, I chanced to sit next to Doctor Cooper.[10] I entertained him during the whole of the time we were together with an account of Arnold's collections, the collection I had seen in Europe, the compliments I had heard in France upon the Philosophical Society of Philadelphia, and concluded with proposing that the future legislature of Massachusetts should institute an Academy of Arts and Sciences.

The doctor at first hesitated; thought it would be difficult to find members who would attend to it; but the principal objection was that it would injure Harvard College by setting up a rival to it that might draw the attention and affections of the public in some degree from it. To this I answered, first, that there were certainly men of learning enough that might compose a society sufficiently numerous; and secondly, that instead of being a rival to the university it would be an honor and an advantage to it. That the president and principal professors would, no doubt, be always members of it; and the meetings might be ordered, wholly or in part, at the college and in that room. The doctor at length appeared better satisfied, and I entreated him to propagate the idea and the plan as far and as soon as his discretion would justify. The doctor did accordingly diffuse the project so judiciously and effectually that the first legislature under the new constitution adopted and established it by law. Afterwards, when attending the convention for forming the constitution, I mentioned the subject to several of the members, and when I was appointed by the subcommittee to make a draft of a project of a constitution to be laid before the convention, my mind and heart was so full of this subject that I inserted the provision for the encouragement of literature in chapter 5, section 2. I was somewhat apprehensive that criticism and objections would be made to the section, and particularly that the "natural history" and the "good humor" would be stricken out; but the whole was received very kindly, and passed the convention unanimously, without amendment.[11]

The two societies are still institutions of national importance, not only because of a time-honored record and useful work, but on account of important general trusts under their control. Although all their meetings are held in the cities where they were founded, their membership is not localized, and to be a Member of the American Philosophical Society or a Fellow of the American Academy, is an honor highly appreciated by every American scientific man.

The Philosophical Society (founded before the separation of the colonies) copied the Royal Society of Great Britain in its corporate name, as well as in that of its transactions, and in its ideals and methods of work took it for a model.

The American Academy, on the other hand, had its origin "at a time when Britain was regarded as an inveterate enemy and France as a generous patron,"[12] and its founders have placed upon record the statement that it was their intention "to give it the air of France rather than that of England, and to follow the Royal Academy rather than the Royal Society."[13] And so in Boston the academy published Memoirs, while conservative Philadelphia continued to issue Philosophical Transactions.

In time, however, the prejudice against the motherland became less intense, and the academy in Boston followed the general tendency of American scientific workers, which has always been more closely parallel with that of England than that of continental Europe, contrasting strongly with the disposition of modern educational administrators to build after German models.

It would have been strange indeed if the deep-seated sympathy with France which our forefathers cherished had not led to still other attempts to establish organizations after the model of the French Academy of Sciences. The most ambitious of these was in connection with the Academy of Arts and Sciences of the United States of America, whose central seat was to have been in Richmond, Virginia, and whose plan was brought to America in 1788 by the Chevalier Quesnay de Beaurepaire. This project, we are told, had been submitted to the King of France and to the Royal Academy of Science, and had received an unqualified indorsement signed by many eminent men, among others by Lavoisier and Condorcet, as well as a similar paper from the Royal Academy of Paintings and Sculpture signed by Vernet and others. A large sum was subscribed by the wealthy planters of Virginia and by the citizens of Richmond, a building was erected, and one professor, Doctor Jean Rouelle, was appointed, who was also commissioned mineralogist in chief and instructed to make natural-history collections in America and Europe.

The population of Virginia, it proved, was far too scattered and rural to give any chance of success for a project which in its nature was only practicable in a commercial and intellectual metropolis, and the academy died almost before it was born.

"Quesnay's scheme was not altogether chimerical," writes H. B. Adams, "but in the year 1788 France was in no position, financial or social, to push her educational system in Virginia. The year Quesnay's suggestive little tract was published was the year before the French Revolution, in which political maelstrom everything in France went down. . . . If circumstances had favored it, the Academy of the United States of America, established at Richmond, would have become the center of higher education not only for Virginia, but for the whole South, and possibly for a large part of the North, if the academy had been extended, as proposed, to the cities of Baltimore, Philadelphia, and New York. Supported by French capital, to which in large measure we owe the success of our Revolutionary war, strengthened by French prestige, by liberal scientific and artistic associations with Paris, then the intellectual capital of the world, the academy at Richmond might have become an educational stronghold, comparable in some degree to the Jesuit influence in Canada, which has proved more lasting than French dominion, more impregnable than the fortress of Quebec." [14]

A scientific society was organized at Williamsburg during the Revolution, but in those trying times it failed for lack of attention on the part of its founders.

Our forefathers in colonial times had their national universities beyond the sea, and all of the young colonists, who were able to do so, went to Oxford or Cambridge for their classical degrees, and to Edinburgh and London for training in medicine, for admission to the bar, or for clerical orders. Local colleges seemed as unnecessary as did local scientific societies.

Many attempts were made to establish local societies before final results were accomplished, and the beginnings of the national college system had a similar history.

In 1619 the Virginia Company of England made a grant of 10,000 acres of land for "the foundation of a seminary of learning for the English in Virginia," and in the same year the bishops of England, at the suggestion of the King, raised the sum of £1,500 for the encouragement of Indian education in connection with the same foundation. A beginning was made toward the occupation of the land, and George Thorpe, a man of high standing in England, came out to be superintendent of the university, but he and 340 other colonists (including all the tenants of the university) were destroyed by the Indians in the massacre of 1622.

The story of this undertaking is told by Professor H. B. Adams in the

History of the College of William and Mary, in which also is given an account of the Academia Virginiensis et Oxoniensis, which was to have been founded on an island in the Susquehanna River, granted in 1624 for the founding and maintenance of a university, but was suspended on account of the death of its projector, and of King James I, and the fall of the Virginia Company.

Soon after, in 1636, came the foundation of Harvard, then in 1660 William and Mary, Yale in 1701, the College of New Jersey in 1746, the University of Pennsylvania in 1751, Columbia in 1754, Brown in 1764, Dartmouth in 1769, the University of Maryland in 1784, that of North Carolina in 1789–1795, that of Vermont in 1791, and Bowdoin (the college of Maine) in 1794.

When Washington became President, one hundred years ago, there were no scientific foundations within this Republic save the American Academy in Boston; and, in the American Philosophical Society, Bartram's Botanic Garden, the private observatory of Rittenhouse, and Peale's Natural History Museum, Philadelphia.

Washington's own inclinations were all favorable to the progress of science; and Franklin, who would have been Vice-President but for his age and weakness, Adams, the Vice-President, and Jefferson, Secretary of State, were all in thorough sympathy with the desire of their chief to "promote as objects of primary importance institutions for the general diffusion of knowledge." All of them were fellows of the American Philosophical Society, and the President took much interest in its proceedings. The records of the society show that he nominated for foreign membership the Earl of Buchan, president of the Society of Scottish Antiquaries, and Doctor James Anderson.

Washington's mind was scientific in its tendencies, and his letters to the English agriculturists (Young, Sinclair, and Anderson) show him to have been a close student of physical geography and climatology. He sent out with his own hand, while President, a circular letter to the best informed farmers in New York, New Jersey, Pennsylvania, Maryland, and Virginia, and having received a considerable number of answers, prepared a report on the resources of the Middle Atlantic States, which was the first of the kind written in America, and was a worthy beginning of the great library of agricultural science which has since emanated from our Government press.

In a letter to Arthur Young, dated December 5, 1791, he manifested great interest in the Hessian fly, an insect making frightful ravages in the wheat fields of the Middle States, and so much dreaded in Great Britain that the importation of wheat from America was prohibited.[15] It was very possibly by his request that a committee of the Philosophical Society prepared and printed an elaborate and exhaustive report, and since its chairman was Washington's Secretary of State, it was practically a governmental affair, the precursor of subsequent entomological commissions, and of our Department of Economic Entomology.[16]

The interest of Washington in the founding of a national university, as manifested in the provisions of his last will and testament, are familiar to all, and I have been interested to learn that his thoughts were earnestly fixed upon this great project during all the years of the Revolutionary war. It is an inspiring thought that, during the long and doubtful struggle for independence, the leader of the American arms was looking forward to the return of peace, in anticipation of an opportunity to found in a central part of the rising empire an institution for the completing of the education of youths from all parts thereof, where they might at the same time be enabled to free themselves in a proper degree from local prejudices and jealousies.

Samuel Blodget, in his Economica, relates the history of the beginning of a national university.

As the most minute circumstances are sometimes interesting for their relation to great events [he wrote], we relate the first we ever heard of a national university: it was in the camp at Cambridge, in October 1775, when major William Blodget went to the quarters of general *Washington*, to complain of the ruinous state of the colleges, from the conduct of the militia quartered therein. The writer of this being in company with his friend and relation, and hearing general Greene join in lamenting the then ruinous state of the eldest seminary of Massachusetts, observed, *merely to console the company of friends*, that to make amends for these injuries, after our war, he hoped, we should erect a noble national university, at which the youth of all the world might be proud to receive instruction. What was thus pleasantly said, Washington immediately replied to, with that inimitably expressive and truly interesting look for which he was sometimes so remarkable: "*Young man you are a prophet! inspired to speak what I feel confident will one day be realized.*" He then detailed to the company his impressions, that all North America would one day become united; he said, that a colonel Byrd,[17] of Virginia, he believed, was the first man who had pointed out the best central seat [for the capital city], *near to the present spot*, or about the falls of the Potomack. General Washington further said, that a Mr. Evans[18] had expressed the same opinion, with many other gentlemen, who from a cursory view of a chart of North America, received this natural and truly correct impression. The look of general Washington, the energy of his mind, his noble and irresistible eloquence, all conspired, so far to impress *the writer* with these subjects, that if ever he should unfortunately become insane, it *will* be from his anxiety for the *federal city* and NATIONAL UNIVERSITY.[19]

In another part of the same book Mr. Blodget describes a conversation with Washington, which took place after the site of the capital had been decided upon, in which the President "stated his opinion, that there were four or five thousand inhabitants in the city of Washington, and until congress were comfortably accommodated, it might be premature to commence a seminary. * * * He did not wish to see the work commenced until the city was prepared for it; but he added, that he hoped he had not omitted to take such measures as would at all events secure the entire object in time, even if its merits should not draw forth from every quarter

the aid it would be found to deserve," alluding, of course, to the provisions in his own will. "He then," continues Blodget, "talked again and
again, on Mr. Turgot's and Doctor Price's calculations of the effect of
compound interest, at which, as he was well versed in figures, he could
acquit himself in a masterly manner."

Concerning the fate of the Potomac Company, a portion of whose
stock was destined by Washington as a nucleus for the endowment of a
university, it is not necessary now to speak. The value of the bequest was
at the time placed at £5,000 sterling, and it was computed by Blodget
that had Congress kept faith with Washington, as well as did the legislature of Virginia in regard to the endowment of Washington College, his
donation at compound interest would in twelve years (1815) have grown
to $50,000, and in twenty-four years (1827) to $100,000, an endowment
sufficient to establish one of the colleges in the proposed university.

Madison, when a member of the Constitutional Convention in 1787,
probably acting in harmony with the wishes of Washington, proposed as
among the powers proper to be added to those of the General Legislature, the following:

To establish a university.

To encourage, by premiums and provisions, the advancement of useful
knowledge and the discussion of science.[20]

That he never lost his interest in the university idea is shown by his
vigorous appeal while President, in his message of December, 1810, in
which he urged the importance of an institution at the capital which
would "contribute not less to strengthen the foundations than to adorn
the structure of our system of government."

Quite in accord with the spirit of Madison's message was a letter in
the Pennsylvania Gazette of 1788,[21] in which it was argued that the new
form of government proposed by the framers of the Constitution could

not succeed in a republic, unless the people were prepared for it by an education adapted to the new and peculiar situation of the country, the most essential instrument for which should be a Federal university. Indeed, the tone of this article, to which my attention has recently been directed by President Welling, was so harmonious with that of the previous and subsequent utterances of Madison as to suggest the idea that he, at that time a resident of Philadelphia, may have been its author. It is more probably, however, that the writer was Benjamin Rush, who in 1787 issued an Address to the people of the United States,[22] which began with the remark that there is nothing more common than to confound the terms of American Revolution with those of the late American war.

"The American war is over," he said, "but this is far from being the case with the American Revolution. On the contrary, nothing but the first act of the great drama is closed. It remains yet to establish and perfect our new forms of government, and to prepare the principles, morals, and manners of our citizens for these forms of government after they are established and brought to perfection."[23]

And then he went on to propose a plan for a national university, of the broadest scope, with post-graduate scholarships, a corps of traveling correspondents, or fellows, in connection with the consular service, and an educated civil service, organized in connection with the university work.

In Economica, the work just quoted, printed in 1806, the first work on political economy written in America, Blodget referred to the national university project as an accepted idea, held in temporary abeyance by legislative delays.

Blodget urged upon Congress various projects which he thought to be of national importance, and among the first of these was To erect, or at least to point out, the place for the statue of 1783, and either to direct or permit the colleges of the university formed by Washington to commence around this statue after the manner of the Timoleonteon of Syracuse.[24]

In intimate connection with his plan for a university was that of Washington for a military academy at West Point. He had found during the Revolution a great want of engineers, and this want caused Congress to accept the services of numerous French engineers to aid our country in its struggle for independence.

At the close of the Revolution, Washington lost no time in commending to Virginia the improvement of the Potomac and James rivers, the junction by canal of Chesapeake Bay and Albemarle Sound of North Carolina. He soon after proceeded to New York to see the plans of General Schuyler to unite the Mohawk with the waters of Lake Ontario, and to Massachusetts to see the plans of the Merrimac Navigation Company.

It was the want of educated engineers for work of this kind that induced Generals Washington, Lee, and Huntington and Colonel Pickering, in the year 1783, to select West Point as a suitable site for a military academy, and at that place such an institution was essayed, under the law of Congress, in 1794. But from the destruction of the building and its contained books and apparatus by fire, the academy was suspended until the year 1801, when Mr. Jefferson renewed the action of the law, and the following year, 1802, a United States Corps of Engineers and Military Academy was organized by law and established at West Point, with General Jonathan Williams, the nephew of Franklin and one of the vice-presidents of the Philosophical Society, at its head, and the United States Military Philosophical Society was established with the whole Engineer Corps of the Army for a nucleus.

This society had for its object "the collecting and disseminating of military science." Its membership during the ten years of its existence included most of the leading men in the country, civilians as well as officers in the Army and Navy. Meetings were held in New York and Washington, as well as in West Point, and it seems to have been the first national scientific society.[25]

The Patent Office also began under Washington, the first American patent system having been founded by act of Congress April 10, 1790.

On the 8th of January, 1790, President Washington entered the Senate Chamber, where both Houses of Congress were assembled, and addressed them on the state of the new nation. In the speech of a few minutes, which thus constituted the first annual message to Congress, about a third of the space was given to the promotion of intellectual objects—science, literature, and arts. The following expression may perhaps be regarded as the practical origination of our patent system:

I can not forbear intimating to you the expediency of giving effectual encouragement, as well to the introduction of new and useful inventions from abroad, as to the exertions of skill and genius in producing them at home.

This, of course, was in direct pursuance of the constitutional enactment, bethought and inserted toward the closing days of the convention in September, 1787, empowering Congress with such authority. Each House, the Senate on the 11th and the Representatives on the 12th, sent a cordial response to the President's address, reciting the particulars of his discourse, and promising, especially to his suggestions for encouragement of science and arts, "such early attention as their respective importance requires" and the lower House proceeded rapidly with the work. January 15 it was resolved that the various measures indicated by the President should be referred to select committees, respectively, and on the 25th such a committee was formed to consider the encouragement of the useful arts. It consisted of Edanus Burke, of South Carolina, a justice of the supreme court of that State, and native of Ireland; Benjamin Huntington, of Connecticut, and Lambert Cadawalader, of New Jersey. On the 16th of February Mr. Burke reported his bill, which passed to its second reading the following day. It was copiously discussed and amended in Committee of the Whole, particularly March 4, when "the clause which gives a party a right to appeal to a jury from a decision of referees, it was moved should be struck out." After a good deal of pointed and profitable remark as to the true sphere and function of juries the motion for striking out was carried.

The next day, March 5, the bill was ordered to be engrossed, and on the 10th, after third reading, it passed and was carried to the Senate. Here, in a few days, it was referred to a committee of which Charles Carroll, of Maryland, was chairman, and reported back the 29th of March, where it passed, with twelve amendments, on the 30th. On the 8th of April it went forward with the signatures of Speaker and Vice-President to the President, who approved it April 10, 1790.[26] The first patent was granted on the 31st of the following July to Samuel Hopkins, of Vermont, for making pot and pearl ashes; and two more during that year.[27]

Thomas Jefferson, Secretary of State at this period, under which Department especially the patent system grew up for more than half its first century, took so keen an interest in its aim and workings, and gave such searching personal attention to the issue of the several patents, that he has been quite naturally reputed as the father of our Patent Office, and it seems to have been supposed that the bill itself creating it proceeded from his own suggestion. But by a comparison of dates this appears hardly possible. Jefferson returned from Europe to Norfolk and Monticello toward the end of 1789, his mind deeply occupied with the stirring movements of revolution abroad. During the winter months he was debating whether he should accept the charge of the State Department, offered him by Washington; making his way by slow stages from Virginia to New York; receiving innumerable ovations; paying his last visit to the dying Franklin, and he only reached the seat of government March 21, when the legislative work on this act was practically finished. More than to any other individual, probably, the American patent system looks for its origin to the Father of the Country.[28]

Jefferson took great pride in it, and gave personal consideration to every application that was made for patents during the years between 1790 and 1793, while the power of revision and rejection granted by that act remained in force. It is a matter of tradition, handed down to us from generation to generation, that when an application for a patent was made he would summon Mr. Henry Knox, of Massachusetts, who was Secretary of War, and Mr. Edmund Randolph, of Virginia, who was Attorney-General, these officials being designated by the act, with the Secretary of State, a tribunal to examine and grant patents; and that these three distinguished officials would examine the application critically, scrutinizing each point of the specification and claims carefully and vigorously. The result of this examination was that, during the first year, a majority of the applications failed to pass the ordeal, and only three patents were granted. Every step in the issuing of a patent was taken with great care and caution, Mr. Jefferson thinking always to impress upon the minds of his officers and the public that it was a matter of no ordinary importance.

The subsequent history of the office is very interesting, especially since it contains a record of Mr. Jefferson's vigorous opposition to the change effected by the act of 1793, which, he held, by a promiscuous granting of exclusive privileges would lead to the creation of monopoly in the arts and industries, and was against the theory of a popular government, and would be pernicious in its effects.

In 1812 a building was put up for the accommodation of the office, but this was destroyed in 1836, and with it most of the records which would be necessary for a proper understanding of the early history of American invention.

In the Patent Office building, and with it destroyed, there was gathered a collection of models, which was sometimes by courtesy called the

American Museum of Art, and which afforded a precedent for the larger collection of models and natural products, which remained under the custody of the Commissioner of Patents until 1858, when it was transferred to the Smithsonian Institution, and became a part of the present National Museum.

In 1836 the patent system was reorganized, and most of the methods at present in use were put in operation. As it now stands, it is one of the most perfect and effective in the world, and the Patent Office, judged by the character of the work it performs, although, perhaps, not strictly to be classed among the scientific institutions, is nevertheless entitled to such a place by reason of its large and admirable corps of trained scientific experts serving on the staff of examiners.[29]

The Administration of John Adams, beginning in 1797, was short and turbulent. Political strife prevented him from making any impression upon our scientific history; but it requires no research to discern the attitude of the man who founded the American Academy and who drew up the articles for the encouragement of literature and science in the constitution of Massachusetts.

Jefferson, as Vice-President, taking little part in the affairs of the Administration, was at liberty to cultivate the sciences. When he came to Philadelphia to be inaugurated Vice-President, he brought with him a collection of the fossilized bones of some large quadruped, and the manuscript of a memoir upon them, which he read before the American Philosophical Society, of which he had been elected president the preceding year.

"The spectacle of an American statesman coming to take part as a central figure in the greatest political ceremony of our country and bringing with him an original contribution to science is certainly," as Luther has said, "one we shall not soon see repeated."[30]

In 1801 began the Administration most memorable in the history of American science. The President of the United States was, during the eight years of his office, president of the American Philosophical Society as well, and was in touch with all the intellectual activities of the period. He wrote to a correspondent, "Nature intended me for the tranquil pursuits of science by rendering them my supreme delight;" and to another he said, "Your first letter gives me information in the line of natural history, and the second promises political news; the first is my passion, the last is my duty, and therefore both desirable."

"At times of the fiercest party conflict," says Luther, "when less happily constituted minds would scarcely have been able to attend to the routine duties of life, we find him yielding to that subtle native force which all through life was constantly drawing him away from politics to science."

Thus, during these exciting weeks in February, 1801, when Congress was vainly trying to untangle the difficulties arising from the tie vote between Jefferson and Burr, when every politician at the capital was busy with schemes and counterschemes, this man, whose political fate was balanced on a razor's edge, was corresponding with Doctor Wistar in regard to some bones of the mammoth which he had just procured from Shawangunk, in New York. Again, in 1808, when the excitement over the Embargo was highest, and when every day brought fresh denunciations of him and his policy, he was carrying on his geological studies in the White House itself. Under his direction upward of 300 specimens of fossil bones had been brought from the famous Big Bone Lick and spread in one of the large unfinished rooms of the Presidential Mansion. Doctor Wistar was asked to come to Philadelphia and select such as were needed to complete the collection of the Philosophical Society. The exploration of the lick was made at the private expense of Jefferson through the agency of General William Clarke, the western explorer, and this may fairly be regarded as the beginning of American governmental work in paleontology.

His scientific tendencies led to much criticism, of which the well-known lines by William Cullen Bryant, in The Embargo, afford a very mild example.[31] He cast all calumny aside with the remark "that he who had nothing to conceal from the press had nothing to fear from it," and calmly went on his way. The senior members of his Cabinet were James Madison, a man of the most enlightened sympathy with science, and Gallatin, one of the earliest American philologists; while one of his strongest supporters in Congress was Samuel Latham Mitchill, a mighty promoter of scientific interests in his native State, whom Adams wittily describes as "chemist, botanist, naturalist, physician, and politician, who supported the Republican party because Jefferson was its leader, and Jefferson because he was a philosopher."

During this administration the project for a great national institution of learning was revived by Joel Barlow. In 1800, when Barlow was the American minister in Paris, he said in a letter to Senator Baldwin:

I have been writing a long letter to Jefferson on quite another subject. . . . It is about learned societies, universities, public instruction, and the advantages you now have for doing something great and good if you will take it up on proper principles. If you will put me at the head of the Institution there proposed, and give it that support which you ought to do, you can't imagine what a garden it would make of the United States: I have great projects, and only want the time and means for carrying them into effect.[32]

M. Dupont de Nemours was also corresponding with Jefferson upon the same subject, and his work, Sur l'Éducation Nationale dans les États-Unis, published in Paris in 1800, was written at his request.[33]

Barlow returned to the American States in 1805, and almost his first public act after his arrival, we are told, was to issue a prospectus in which he forcibly and eloquently depicted the necessity and advantages of a national scientific institution.

This was to consist of a central university at or near the seat of government, and, as far as might seem practicable or advisable, other universities, colleges, and schools of education, either in Washington or in other parts of the United States, together with printing presses for the use of the institution, laboratories, libraries, and apparatus for the sciences and the arts, and gardens for botany and agricultural experiments.

The institution was to encourage science by all means in its power, by correspondence, by premiums and by scholarships, and to publish schoolbooks at cost of printing.

The Military and Naval Academies, the Mint, and the Patent Office were to be connected with the university, and there was also to be a general depository of the results of scientific research and of the discoveries by voyages and travels, actually the equivalent of a national museum.

"In short," wrote Barlow, "no rudiment of knowledge should be below its attention, no height of improvement above its ambition, no corner of an empire beyond its vigilant activity for collecting and diffusing information." [34]

The editor of the National Intelligencer, the organ of the Administration in 1806, commented favorably upon the plan of Barlow.

This gentleman [he wrote], whose mind has been enlarged by extensive observation, by contemplating man under almost every variety of aspect in which he appears, and whose sentiments have been characterized by an uniformly zealous devotion to liberty, has most justly embraced the opinion that the duration as well as perfection of republicanism in this country will depend upon the prevalence of correct information, itself dependent upon the education of the great body of the people. Having raised himself, as we understand, to a state of pecuniary independence, he has returned to his native country, with a determination of devoting his whole attention and labors to those objects which are best calculated to improve its state of society, its science, literature, and education. The disinterested exertions of such a man merit the national attention. [35]

Barlow's prospectus, we are told, was circulated throughout the country, and met with so favorable a response that in 1806 he drew up a bill for the incorporation of the institution, which Mr. Logan, of Philadelphia, introduced in the Senate, which passed to a second reading, was referred to a committee which never reported, and so was lost.

Barlow's National Institution resembled more closely the House of Salomon in The New Atlantis of Bacon than it did the eminently practical university project of Washington. It would be interesting to know to what extent President Jefferson was in sympathy with Barlow. The mind which a few years later directed the organization of the University of Virginia could scarcely have approved all the features of the Kalorama plan. He was undoubtedly at this time anxious that a national university should be founded, as is shown by his messages to Congress in 1806 and 1808,[36] though it is probable that he wished it to be erected in some convenient part of Virginia, rather than in the city of Washington. The project for transplanting to America the faculty of the College of Geneva, which, but for the opposition of Washington, would probably have been attempted in 1794, had reference rather to the formation of a State university, national in influence, than to a central Federal institution.[37]

Although Barlow's plan was, in its details, much too elaborate for the times, the fundamental ideas were exceedingly attractive, and led to very important and far-reaching results.

Barlow expected, of course, that his institution should be established and maintained at Government cost. This was soon found to be impracticable, and those who were interested in the intellectual advancement of the capital soon had recourse to the idea of beginning the work at private expense, relying upon Government aid for its future advancement.

Barlow's classmate, Josiah Meigs, his friend and neighbor Thomas Law, aided by Edward Cutbush, Judge Cranch, and other citizens of Washington, proceeded forthwith to attempt that which the politicians dared not.

The essential features of Barlow's plan were:

(1) The advancement of knowledge by association of scientific men; and

(2) The dissemination of its rudiments by the instruction of youth.[38]

To meet the first of these requirements they organized the Columbian Institute for the Promotion of Arts and Sciences, in 1819; and for the second, the Columbian College, incorporated in 1821. Most of the prominent members of the Columbian Institute were also among the friends and supporters of the college. Doctor Josiah Meigs, the friend and classmate of Barlow, the president of the institute from 1819 to 1821, was an incorporator and a member of the first faculty of the college.[39]

Doctor Edward Cutbush, the founder of the Columbian Institute, was also a professor, as well as Doctor Thomas Sewall, Doctor Alexander McWilliams, and Judge William Cranch, and in publications made at the time these men distinctly proposed to realize the aspirations of Washington for the creation of a great national university at the seat of the Federal Government. It was in this cause President Monroe gave to the Columbian College his public support as President of the United States. At a later day, when an hour of need overtook the college, John Quincy Adams became one of its saving benefactors.[40]

The donation of $25,000 made to the Columbian College in 1832 was preceded by a report from the Committee in House of Representatives on the District of Columbia.

That report may be found in Reports of committees, first session Twenty-second Congress (1831–32), III, Report No. 334.

After reciting the early history of the college the report proceeds as follows:

Few institutions present as strong claims to the patronage of Government, as that, in behalf of which the forementioned memorial has been presented. [The report is made in answer to a memorial of the president and trustees of the college, asking Congress to make a donation to the college 'from the sale of public lots or from such other source as Congress may think proper to direct.'] Its location near the seat of Government, its salubrious middle climate, and other advantages, and the commendable efforts of its present trustees and professors to sustain it, justly entitle it to public benificence.

The Columbian Institute was granted the use of rooms in the Capitol building under the present Congressional Library Hall, which became a center of the scientific and literary interests of Washington, and its annual meetings were held in the Hall of the House of Representatives, where Southard, Clay, Everett, Meigs, and Adams delivered addresses upon matters of science and political economy to large assemblages of public men. In 1819, Josiah Meigs, its president, writing to Doctor Daniel Drake, of Cincinnati, said:

I have little doubt that this Congress will, before they rise, give the Institute a few acres of ground for our building and for a Botanic Garden. Mr. Barlow made great efforts to obtain this object eight or ten years ago—he could do nothing—but prejudices which *then* were of the *density* of thunder cloud are now as *tenuous* as the tail of a Comet.[41]

The supreme legislative power of the United States over persons and property within the District of Columbia, is unquestioned. Congress has repeatedly made grants of portions of the public lands to seminaries of learning situated within the limits of States and Territories, where such lands lie. The constitution having thus confided to the care of the National Legislature, the rights and interests of the people of the District of Columbia, and Congress having made liberal donations out of the national domain to promote the great cause of education in all the other districts within which the General Government has exclusive jurisdiction, it would seem to be cruel injustice to refuse the small boon now recommended. These considerations, induce the hope that the proposed donation will be exempt from all opposition, not founded in doubts of the just claim to patronage of the institution for the benefit of which it is designed. And these claims, it is fully believed, will stand the test of the severest scrutiny.

The report from which the above exacts are taken was made February 27, 1832 (to accompany House bill No. 422), by Mr. Thomas, of Maryland (on behalf of the Committee on the District of Columbia), in answer to memorial of the trustees and the president of the Columbian College.

On the ground granted by Congress, a botanical garden was established by the society in 1822 or 1823 with the cooperation of the State Department and the consular service. In 1829 the society applied to Congress for pecuniary aid, which was not granted.[42]

The Columbian University was also an applicant for Government aid, which it received to the amount of $25,000 in 1832, on the ground that it was an institution of national importance, organized by private individuals to do work legitimately within the domain of governmental responsibilities.[43]

The Columbian College received nearly one-third of its original endowment from the Government of the United States. Of the remainder, perhaps one-half was contributed by men like President Adams, whose sole interest in it was a patriotic one.

During Jackson's Presidency all ideas of centralization, even in scientific matters, appear to have fallen into disfavor, and the Columbian Institute and the Columbian College were forced to abandon their hopes for governmental aid. The institute languished and dropped out of existence, while the college, under the fostering care of a church organization (which finally dropped it in 1846), and through the beneficence of individuals, one of whom, a citizen of Washington, gave it property to the value of $200,000, has grown to be a university in name and scope, and is included among the thirteen "foundations comprising groups of related faculties, colleges, or schools," enumerated in the Report of the Commissioner of Education for 1886–87.

Although it has not since 1832 made any claims for Government aid, nor assumed to be in any way a ward of the nation, its early history is significant, on account of its connection with the project for a national university, which has been for more than a century before the people. The Government has since established in Washington City the National Deaf-Mute College, which it still maintains, and the Howard University, intended primarily for the freedman but open to all.

The founders of the Columbian Institute and the Columbian University were building better than they knew, for they were not only advancing knowledge in their own day and generation, but they were educating public opinion for a great opportunity, which soon came in the form of a gift to the nation from beyond the sea in the form of the Smithson bequest.

The story of the Smithsonian Institution is a remarkable one. Smithson was a graduate of the University of Oxford, a fellow of the Royal Society, a chemist and mineralogist of well-recognized position. The friend and associate of many of the leading scientific men of England, he found it advisable, for reasons connected with his family history, to pass most of his life upon the Continent. A man of ample fortune, he associated with men of similar tastes, and died in 1829, leaving in trust to the United States property now amounting in value to nearly three-quarters of a million of dollars to establish at the national capital "an institution for the increase and diffusion of knowledge among men." No one has been able to explain why he did this. He had, so far as we know, no friend or correspondent in the United States, and had made known to no one his intention of establishing an institution of learning in the New World.[44]

It is more than probable, however, that he knew Barlow when American minister in Paris, and that the prospectus of the National Institution or the treatise by Dupont de Nemours may have attracted his attention. He was aware of the failure of the attempts to obtain national support at the start for scientific uses, and conceived the idea of founding, with his own means, an organization which should, he foresaw, grow into national importance. Anyone who will take the pains to compare the criticisms and objections to Barlow's project, as set forth in Wirt's essay in The Old Bachelor,[45] with those which were urged in Congress and the public press in opposition to the acceptance of the Smithson bequest thirty years later, can not fail to be greatly impressed by the similarity of tone and argument.

The Smithsonian Institution, with its dependencies and affiliations, corresponds perhaps more closely at the present time to Barlow's National Institution than any organization existing elsewhere in the world. The names of its three secretaries—Henry, the physicist (in office from 1846 to 1878); Baird, the naturalist (Assistant Secretary from 1850 to 1878, Secretary, 1878–1887); and Langley, the astronomer, suggest in a few words the main features of its history.

Recurring to Jefferson's presidency, it should be noted that its most important scientific features were the inception of the system of scientific surveys of the public domain, and the organization of the Coast Survey. The first was most peculiarly Jefferson's own, and was the outcome of more than twenty years of earnest endeavor.

The apathy of the British Government in colonial times in the matter of explorations of the American continent is inexplicable. Halley, the philosopher and mathematician, was in charge of a fruitless expedition in 1699; and Ellis, in 1746, explored Hudson Bay under Government auspices, searching for a northwest passage.

The first inland exploring expedition under Government auspices seems to have been that of Governor Spotswood, of Virginia, who in 1724, accompanied by a party of young colonists, made an excursion to the summit of the Blue Ridge for the purpose of ascertaining what lay beyond.

Nothing else was done in colonial days, although it would appear that Jefferson, and doubtless others as well as he, had in mind the importance of exploring the great Northwest. In the recently published life of Matthew Fontaine Maury, the story is told of his grandfather, the Rev. James

Maury, an Episcopal clergyman and instructor of youth in Walker parish, Albemarle County, Virginia, who numbered among his pupils three boys who afterwards became Presidents of the United States and five signers of the Declaration of Independence. He was a quiet thinker—a serene old man who gave the week to contemplative thought and to his school, and Sunday to the service of the sanctuary. In 1756 he was already dazzled by the rising glory of the new country. He was intensely interested in the great Northwest. The Missouri was a myth at that time. Cox had ascended the Mississippi to the falls of St. Anthony, and reported the existence of such a stream, but all beyond was shrouded in mystery.

"But see," said the aged clergyman, pointing with trembling finger and eager eye to the map of the North American Continent—"see, there must be a large river in that direction: mountains are there, and beyond them there must be a stream to correspond with the vast river on this side of the chain." And by a process of reasoning based on physical geography, he pointed out to his pupils (Thomas Jefferson among them) the existence and line of the river as accurately as Le Verrier did the place of Neptune in the firmament, and predicted that a great highway to the West would some day be opened in this direction.[46]

It would appear that Jefferson never forgot the suggestion of his venerable teacher. While minister of the United States in Paris, in 1785, he became acquainted with John Ledyard, of Connecticut, a man of genius, of some science, and of fearless courage and enterprise, who had accompanied Captain Cook on his voyage to the Pacific. "I suggested to him," writes Jefferson, "the enterprise of exploring the western part of our continent by passing through St. Petersburg to Kamchatka, and procuring a passage thence in some of the Russian vessels to Nootka Sound, whence he might make his way across the continent to the United States." He proceeded to within 200 miles of Kamchatka, and was there obliged to take up his winter quarters, and when preparing in the spring to resume his journey, he was arrested by an officer of the Empress of Russia, and carried back in a closed carriage to Poland. "Thus," says Jefferson, "failed the first attempt to explore the western part of our northern continent."

In a letter to Bishop Madison, dated Paris, July 19, 1788, Jefferson tells the story of Ledyard's failure, and of his departure on an expedition up the Nile. "He promises me," continues Jefferson, "if he escapes through his journey, he will go to Kentucky and endeavor to penetrate westwardly to the South Sea." Ledyard died in Africa.

The proposed expedition of Ledyard, though undertaken at the instance of the American minister in Paris, can scarcely be regarded as a governmental effort. It is of interest, however, as leading up to the second attempt, which also was inspired and placed on foot by Jefferson.

In 1792, [writes Jefferson,] I proposed to the American Philosophical Society, that we should set on foot a subscription to engage some competent persons to explore those regions in the opposite direction—that is, by ascending the Missouri, crossing the Stony Mountains, and descending the nearest river to the Pacific.[47]

Captain Meriwether Lewis, being then stationed at Charlottesville on the recruiting service, warmly solicited me to obtain for him the execution of that object. I told him that it was proposed that the person engaged should be attended by a single companion only, to avoid exciting alarm among the Indians. This did not deter him, but Mr. Andre Michaux, a professed botanist, author of the Flora Boreali-Americana, and of the Histoire des Chênes de l'Amérique, offering his services, they were accepted. He received his instructions, and when he had reached Kentucky in the prosecution of his journey he was overtaken by an order from the minister of France, then at Philadelphia, to relinquish the expedition and to pursue elsewhere the botanical inquiries on which he was employed by the Government, and thus failed the second attempt to explore that region.[48]

It is related by Jefferson, in his Memoranda of Conversations, that Judge Breckenridge, of Kentucky, told him in 1800, that Michaux was not only a botanical agent of the French, but a political emissary, and that he held a commission as commissary for an expedition against the Spaniards, planned by Genet, in connection with a plot to gain possession of the eastern Mississippi Valley for France.[49]

In 1803, [continues Jefferson] the act of establishing trading houses with the Indian tribes being about to expire, some modifications of it were recommended to Congress by a confidential message of January 18, and an extension of its views to the Indians on the Missouri. In order to prepare the way, the message proposed sending an exploring party to trace the Missouri to its source, to cross the highlands, and follow the best water communication which offered itself from thence to the Pacific Ocean. Congress approved the proposition and voted a sum of money for carrying it into execution. Captain Lewis, who had then been near two years with me as private secretary, immediately renewed his solicitation to have the direction of the party.

In his life of Lewis, prefixed to the history of the expedition, Jefferson gives in full an account of Lewis's preparation for the expedition, including his instruction in astronomical observation by Andrew Ellicott, and also a full text of the instructions, signed by him, addressed to Lewis and his associate, Captain William Clarke. Captain Lewis left Washington on the 5th of July, 1803, and proceeded to Pittsburg. Delays of preparation, difficulties of navigation down the Ohio, and other obstructions retarded his arrival at Cahoki until the season was so far advanced that he was obliged to wait until the ice should break up in the beginning of spring. His mission accomplished, he returned to St. Louis on the 23d of September, 1806.

Never, [says Jefferson,] did a similar event excite more joy through the United States. The humblest of its citizens had taken a lively interest in the issue of the journey, and looked forward with impatience for the information it would furnish. The anxiety, too, for the safety of the corps had been kept in a state of excitement by lugubrious rumors circulated from time to time on uncertain authorities, and uncontradicted by letters or other direct information, from the time they had left the Mandan towns on their ascent up the river in April of the preceding year, 1805, until their actual return to St. Louis.

The second expedition toward the West was also sent out during Jefferson's Administration, being that under the command of General Zebulon M. Pike, who was sent to explore the sources of the Mississippi River and the western parts of Louisiana, continuing as far west as Pikes Peak, the name of which still remains as a memorial of this enterprise.[50]

The expedition of Lewis and Clarke was followed in due course and in rapid succession by others, some geographical, some geological, some for special researches, and some more comprehensive in character.

To those who are in the least degree familiar with the history of American exploration the names of Long, Cass and Schoolcraft, Bonneville, Nicollet, Frémont, Sitgreaves, Wizlizenus, Foster and Whitney, Owen, Stansbury, Abert, Marcy, Stevens, Gunnison, Beckwith, Whipple, Williamson, Parke, Pope, Emory, Bartlett, Bryan, Magraw, Johnston, Campbell, Warren, Twining, Ives, Beale, Simpson, Lander, McClellan, Mullan, Raynolds, Heap, Jones, Ruffner, Ludlow, Maguire, Macomb, and Stone will bring up the memory of much adventurous exploration and a vast amount of good scientific work; while to mention Hayden, Wheeler, King, and Powell is to leave the field of history and to call up the early stages of the development of that magnificent organization, the United States Geological Survey, which is still in the beginning of its career of usefulness.[51]

The history of the Coast Survey began with the earliest years of the century. It has been thought by some that the idea originated with Albert Gallatin, and by others that it was due to Professor Robert Patterson,[52] while Hassler, whose name is so intimately associated with its early history, seems to have supposed that it was suggested by his own advent, in 1805, bringing with him from Switzerland a collection of mathematical books and instruments.[53]

Passing by the question as to who was the originator of the idea, with the simple remark that it is doubtful whether such an enterprise should not have for long years been in the minds of many Americans, it may be said that, without doubt, the early organization of the survey was due to the scientific wisdom and political foresight of Jefferson, who realized that within a few years the country would be involved in a war with Great Britain, and that a thorough knowledge of the coast was essential, not only to the prosperity of the nation in time of peace, but still more to its safety in case of invasion. At that time the only charts available for our mariners were those in The Atlantic Neptune of Colonel Des Barres, and the old hydrographic charts issued by the Dutch, French, and English Governments. Jefferson realized that American seamen were less familiar with many portions of their own coast than were the European navigators, and he appreciated fully the importance of having a knowledge of this kind far more accurate than that which was possessed by any foreigner. "With the clear and bold perception which always distinguishes men of genius when they are trusted in times of danger with the destiny of nations, the President recommended the survey of the home coast with all the aid of the more recent discoveries in science;" and in his annual message to Congress, in the year 1807, proposed the establishment of a national survey, for the purpose of making a complete chart of the coast with the adjacent shoals and soundings.

In response to this recommendation, Congress made an appropriation of $50,000 for the purpose of carrying out the provision of the following law:

AN ACT to provide for surveying the Coasts of the United States.

Be it enacted, etc., That the President of the United States shall be, and he is hereby, authorized and requested to cause a survey to be taken of the coasts of the United States, in which shall be designated the islands and shoals, with the roads or places of anchorage, within twenty leagues of any part of the shores of the United States; and also the respective courses and distances between the principal capes, or head lands, together with such other matters as he may deem proper for completing an accurate chart of every part of the coasts within the extent aforesaid. (Act of February 10, 1807.)

By the direction of the President, Albert Gallatin, Secretary of the Treasury, addressed a circular letter to American men of science, requesting their opinion as to the character of the plan to be adopted.

In the circular of the Secretary of the Treasury, the work to be performed was defined as consisting of three distinct parts, as follows:

(1) The ascertainment by a series of astronomical observations of the position of a few remarkable points on the coast, and some of the light-houses placed on the principal capes, or at the entrance of the principal harbors, appear to be the most eligible places for that purpose, as being objects particularly interesting to navigators, visible at a great distance, and generally erected on spots on which similar buildings will be continued so long as navigation exists.

(2) A trigonometrical survey of the coast between those points of which the positions shall have been astronomically ascertained; in the execution of which survey, the position of every distinguishable permanent object should be carefully designated; and temporary beacons be erected at proper distances on those parts of the coast on which such objects are really to be found.

(3) A nautical survey of the shoals and soundings of the coast, of which the trignonometrical survey of the coast itself, and the ascertained position of the light-houses, and other distinguishable objects, would be the basis; and which would therefore depend but little on any astronomical observations made on board the vessels employed on that part of the work.

This circular letter was submitted to thirteen scientific men, and in response thirteen plans were received at the Treasury Department. A commission, composed of the experts from whom answers had been received, was formed. They met at Professor Patterson's, in Philadelphia, and the plan which they finally selected was then proposed by Ferdinand Rudolph Hassler, at that time, and for several years thereafter, professor in the Military Academy at West Point.

Nothing was done to secure definitely the execution of this plan until 1811, when Hassler was sent to Europe to procure the necessary instruments and standards of measure for the proposed work. He was detained as an alien in London during the entire war with England, and until 1815, when he returned to the United States, having, as a matter of course, far exceeded the limits of his appropriation, with a large claim against the Government for indemnification. [54]

I have been unable to ascertain the exact date of the appointment of Hassler as the Superintendent of the Coast Survey, although it was thoroughly understood at the time of the acceptance of his plan in 1807 that it was to be carried out under his direction.

It was not until August, 1816, that the contract was signed with the Government which authorized Hassler to proceed with his work. In 1817 a beginning was made in the bay and harbor of New York, but Congress failed to provide for its continuance, and it was soon suspended, and in 1818, before the Superintendent had the opportunity to publish a report upon the results of his last year's labor, Congress, on the plea "that the little progress hitherto made in the work had caused general dissatisfaction," ordered its discontinuance by repealing the law under which the Superintendent had been appointed, and providing that no one should be employed in the survey of the coast except officers of the Army and Navy. This was practically a discontinuance of the work, because there was no one in America but Hassler who was capable of directing it.

Immediately after being thus legislated out of office he was appointed one of the astronomers to represent the United States in the settlement of the Canadian boundary.

From 1819 to 1832 attempts were made at various times by the Navy Department to survey several portions of the coast. A few detached surveys were made, but no general systematic work was attempted, and the result was not on the whole creditable. In 1828 the Hon. S. L. Southard, of New Jersey, at that time Secretary of the Navy, in response to resolutions of inquiry from the House of Representatives, admitted that the charts produced by the Navy were unreliable and unnecessarily expensive, and declaring also that the plan which had been employed was desultory and unproductive, recommended that the provisions of the law of 1807 should be resumed.

In 1832 Congress passed an act reorganizing the surveys on the old plan.

AN ACT to carry into effect the act to provide for a survey of
the Coasts of the United States

[SEC. 1.] *Be it enacted, etc.*, That for carrying into effect the act entitled "An act to provide for surveying the coasts of the United States," approved on the 10th day of February, 1807, there shall be, and hereby is, appropriated a sum not exceeding twenty thousand dollars, to be paid out of any money in the Treasury not otherwise appropriated; and the said act is hereby revived, and shall be deemed to provide for the survey of the coasts of Florida in the same manner as if the same had been named therein.

[SEC. 2.] That the President of the United States be, and he is hereby, authorized, in and about the execution of the said act, to use all maps, charts, books, instruments, and apparatus, which now or hereafter may belong to the United States, and employ all persons in the land and naval service of the United States, and such astronomers and other persons as he shall deem proper.

Hassler was now again appointed Superintendent of the Coast Survey, and held his position until his death in 1843, the work for a short time, at first, being assigned to the Treasury Department, and in 1834 transferred to the Navy Department, and in 1836 again retransferred to the

Treasury, where it has since remained, its status being finally definitely settled by act of Congress passed in 1843, shortly before the appointment of Alexander Dallas Bache, as the successor of the first Superintendent of the Survey.

At the time of Hassler's death the survey had been extended from New York, where it was begun, eastward to Point Judith, and southward to Cape Henlopen.

It should be mentioned that in 1825, during the period of the suspension of activity, Hassler presented to the American Philosophical Society a memoir on the subject of the survey, which contained a full account of the plan which he had adopted, a description of his instruments, and a history of what had been accomplished up to 1817. "This memoir," wrote Professor Henry in 1845, "was received with much favor by competent judges abroad, and the commendation bestowed upon it was of no little importance in the wakening of sentiments of national pride, which had considerable influence in assisting the passage of the act authorizing the renewal of the survey in 1832."

With the appointment of Bache as Superintendent in 1843, the Survey entered upon a new period of prosperity, the discussion of which is not within the province of this paper, and it seems appropriate to close this notice of the origin and early history of the organization by quoting from the first report of his successor an estimate of the value of Hassler's services.

The coast survey [wrote Bache] owes its present form, and perhaps its existence, to the zeal and scientific ability of the late superintendent, who devoted the energies of a life to it; and who, but for its interruption at a period when he was in the prime of manhood, and its suspension for nearly fifteen years, might have seen its completion. The difficult task of creating resources of practical science for carrying on such a work upon a suitable scale, required no common zeal and perseverance for its accomplishment, especially at a time (1807) when our country was far from having attained her present position in scientific acquirement, and when public opinion was hardly sufficiently enlightened to see the full advantage of thoroughness in executing the work. For his successful struggle against great difficulties, his adopted country will, no doubt, honor his memory as the pioneer of a useful national undertaking.[55]

The history of the Coast Survey under the successive superintendentships of Bache [1843–1867], Peirce [1867–1874], Patterson [1874–1881], and Hilgard [1881–1887], would make a volume in itself. Under its present Director, Professor Mendenhall, it is growing into renewed vigor and efficiency.

The Coast Survey was the last of the great scientific enterprises begun in Jefferson's Administration. If the Sage of Monticello were now living, what delight he would feel in the manifold scientific activities of the nation. The enlightened policy of our Government in regard to scientific and educational institutions is doubtless to a considerable degree due to his abiding influence.

Nowhere in all the long course of Mr. Jefferson's great career [writes Henry Adams] did he appear to better advantage than when in his message of 1806 he held out to the country and the world that view of his ultimate hopes and aspirations for natural development, which was, as he then trusted, to be his last bequest to mankind. Having now reached the moment when he must formally announce to Congress that the great end of relieving the nation from debt was at length within reach, and with it the duty of establishing true republican government was fulfilled, he paused to ask what use was to be made of the splendid future thus displayed before them. Should they do away with the taxes? Should they apply them to the building up of armies and navies? Both relief from taxation and the means of defense might be sufficiently obtained without exhausting their resources, and still the great interests of humanity might be secured. These great interests were economical and moral; to supply the one, a system of internal improvement should be created commensurate with the magnitude of the country; "by these operations new channels of communication will be opened between the States, the lines of separation will disappear, their interests will be identified, and their union cemented by new and indissoluble ties." To provide for the other, the higher education should be placed among the objects of public care; "a public institution can alone supply those sciences which, though rarely called for, are yet necessary to complete the circle, all the parts of which contribute to the improvement of the country and some of them to its preservation." A national university and a national system of internal improvement were an essential part, and indeed the realization and fruit, of the republican theories which Mr. Jefferson and his associates put in practice as their ideal of government.[56]

Madison's Administration, which began in 1809, though friendly to science, was not characterized by any remarkable advances (except that the Coast Survey was actually organized for work under Hassler, after his return from Europe, in 1816). The war of 1812 and the unsettled state of public affairs were not propitious to the growth of learned institutions.

Monroe became Chief Magistrate in 1817. He, like Madison, was a friend and follower of Jefferson, and in the atmosphere of national prosperity scientific work began to prosper, and there was a great accession of popular interest, and State geological surveys began to come into existence. Schoolcraft and Long led Government expeditions into the West; the American Geological Society and the American Journal of Science were founded.

The city of Washington began to have intellectual interests, and public-spirited men organized the Columbian Institute and the Columbian University.

Monroe was not actually acquainted with science, but was in hearty sympathy with it. When he visited New York, in 1817, he visited the New York Institution, and was received as an honorary member of the Literary and Philosophical Society, and in his reply to the address of Governor Clinton, its president, he remarked that "the honor, glory, and prosperity of the country were intimately connected with its literature and science, and that the promotion of knowledge would always be an object of his attention and solicitude." [57]

The most important new enterprise was in the direction of organizing a national meteorological service.

The first move was made by Josiah Meigs, who was in 1814 appointed Commissioner of the General Land Office. With the exception of Franklin,[58] he was perhaps the earliest scientific meteorologist in America, having, while living in the Bermudas from 1789 to 1794, made a series of observations which he communicated to the Royal Society.[59]

In 1817, or before he began to advocate Congressional action for the establishment of meteorological registers in connection with the Land Office, writing to Doctor Daniel Drake, in 1817, he said:

If my plan be adopted, and the *Registers* be furnished with the requisite Instruments for *Temperature, Pressure, Rain, Wind,* etc., . . . we may in a course of years know more than we shall be able to know on any other plan (p. 82).

Without some system of this kind, our Country may be occupied for ages, and We the people of the United States be as ignorant on this subject as the *Kickapoos* now are, who have occupied a part of it for ages past (p. 82).

In 1817 he also issued a circular to the registrars of the land offices of the several States calling upon them to take regularly certain observations and make monthly official reports upon all meteorological phenomena.

In 1819 a cooperative movement was begun under the direction of Doctor Joseph Lovell, Surgeon-General of the Army, in connection with the medical officers at the principal military posts, by whom reports were made at the end of each month upon the temperature, pressure, and moisture of the air, the amount of rain, the direction and force of the wind, the appearance of the sky, and other phenomena.

The Land Office circular was a remarkable one, and led to the extensive system of Patent Office observations, the results of which, published in connection with those of the War Department and the Smithsonian in 1859, formed the foundation of scientific meteorology in the United States.

In 1839 a most admirable paper by the French geologist, J. N. Nicollet, an Essay on Meteorological Observations, was published under the direction of the Bureau of Topographical Engineers. Some years later the lake system of meteorological observations was established by the Engineer Department, under the direction of Captain (afterwards General) George G. Meade. This included a line of stations extending from the western part of Lake Superior to the eastern part of Lake Ontario.

In 1835 a system of observations had been established under the direction of the board of regents of the University of the State of New York, the points of observation being at the academies of the State; and in 1837 the legislature of Pennsylvania made an appropriation of $4,000 for instruments for use in meteorological observations, which were continued until about 1847. Those of New York were kept up until 1865 or later.

In the meantime the idea of the preannouncement of storms by telegraph was suggested in 1847 by W. C. Redfield, the discoverer of the law of storms, while Lieutenant Maury, from 1851 onward, and especially at the International Meteorological Conference (held at his instance in Belgium in 1853), was promoting the establishment of a system of agricultural meteorology for farmers and of daily weather reports by telegraph.[60]

In February, 1855, Leverrier obtained the sanction of the Emperor of France for the creation of an extensive organization for the purpose of distributing weather intelligence, though it was not till 1860 that he felt justified in making his work international.[61] In 1861 and in 1862 a similar organization was begun in England, under Admiral Fitzroy, which was extended a little later to India.

In the meantime all the essential features for the prediction of meteorological phenomena were in existence in the Smithsonian Institution as early as 1856, having grown up as the result of an extensive series of tabulations of observations recorded by volunteer observers in all parts of the country.

The following historical notes on weather telegraphy, prepared by Professor Cleveland Abbe in 1871,[62] give a summary of the progress of this work:

However frequently the idea may have been suggested of utilizing our knowledge by the employment of the electric telegraph, it is to Professor Henry and his assistants in the Smithsonian Institution that the credit is due of having first actually realized this suggestion.

The practical utilization of the results of scientific study is well known to have been in general greatly furthered by the labors of this noble Institution, and from the very beginning Professor Henry has successfully advocated the feasibility of telegraphic storm warnings. The agitation of this subject in the United States during the years 1830–1855, may be safely presumed to have stimulated the subsequent action of the European meteorologists. It will be interesting to trace the gradual realization of the earlier suggestions of Redfield and Loomis, in the following extracts from the annual Smithsonian Reports of the respective years:

1847: The extended lines of telegraph will furnish a ready means of warning the more northern and eastern observers to be on the watch for the first appearance of an advancing storm.

1848. As a part of the system of meteorology, it is proposed to employ, as far as our funds will permit, the magnetic telegraph in the investigation of atmospherical phenomena. . . . The advantage to agriculture and commerce to be derived from a knowledge of the approach of a storm by means of the telegraph, has been frequently referred to of late in the public journals; and this we think is a subject deserving the attention of the Government.

1849. Successful applications have been made to the presidents of a number of telegraph lines to allow us at a certain period of the day the use of the wires for the transmission of meteorological intelligence. . . . as soon as they [certain instructions, etc.] are completed, the transmission of observations will commence. [It was contemplated to constitute the telegraph operators the observers.]

1850. This map [an outline wall map] is intended to be used for presenting the successive phases of the sky over the whole country at different points of time, as far as reported.

1851. Since the date of the last report the system particularly intended to investigate the nature of American storms immediately under the care of the Institution, has been continued and improved.

The system of weather reports thus inaugurated continued in regular operation until 1861, when the disturbed condition of the country rendered impossible its further continuance. Meanwhile, however, the study of these daily morning reports had led to such a knowledge of the progress of our storms, that in the Report for 1857, Professor Henry writes:

1857. We are indebted to the National Telegraph Line for a series of observations from New Orleans to New York and as far westward as Cincinnati, which have been published in the Evening Star of this city.

We hope in the course of another year to make such an arrangement with the telegraph lines as to be able to give warnings on the eastern coast of the approach of storms, since the investigations which have been made at the Institution fully indicate the fact that as a general rule the storms of our latitude pursue a definite course.

It would seem, therefore, that nothing but the disturbances of the late war prevented our having had ten years ago a valuable system of practical storm warnings. Even before peace had been proclaimed, Professor Henry sought to revive the systematic daily weather reports, and in August, 1864, at the meeting of the North American Telegraph Association (see their published Report of Proceedings), a paper was presented by Professor Baird, on behalf of the Smithsonian Institution, requesting the privilege of the use of the telegraph lines, and more especially in order to enable Professor Henry "to resume and extend the Weather Bulletin, and to give warning of important atmospheric changes to our seaboard." In response to this communication it was resolved, "That this Association recommend . . . . . . to pass free of charge, . . . brief meteorological reports, . . . for the use and benefit of the Institution."

On the communication of this generous response, preparations were at once made for the laborious undertaking, and the inauguration of the enterprise was fixed for the year 1865. In January of that year, however, occurred the disastrous fire which so seriously embarrassed the labors of the Smithsonian Institution for several following years: it became necessary to indefinitely postpone this meteorological work, which indeed had through its whole history been carried on with most limited financial means, and was quite dependent upon the liberal cooperation of the different telegraph companies.

It will thus be seen that without material aid from the Government, but through the enlightened policy of the telegraph companies, and with the assistance of the munificent bequest of James Smithson, "for the increase and diffusion of knowledge," the Smithsonian Institution, first in the world, organized a comprehensive system of telegraphic meteorology, and has thus given first to Europe and Asia, and now to the United States, that most beneficent national application of modern science, the Storm Warnings.

In the report of the Smithsonian Institution for 1858 it is stated:

An object of much interest at the Smithsonian building is a daily exhibition on a large map of the condition of the weather over a considerable portion of the United States. The reports are received about ten o'clock in the morning, and the changes on the maps are made by temporarily attaching to the several stations pieces of card of different colors to denote different conditions of the weather as to clearness, cloudiness, rain, or snow. This map is not only of interest to visitors in exhibiting the kind of weather which their friends at a distance are experiencing, but is also of importance in determining at a glance the probable changes which may soon be expected.[63]

In a still earlier report Professor Henry said:

We are indebted to the National Telegraph line for a series of observations from New Orleans to New York, and as far westward as Cincinnati, Ohio, which have been published in the "Evening Star," of this city. These reports have excited much interest, and could they be extended farther north, and more generally to the westward, they would furnish important information as to the approach of storms. We hope in the course of another year to make such an arrangement with the telegraph lines as to be able to give warning on the eastern coast of the approach of storms, since the investigations which have been made at the Institution fully indicate the fact that as a general rule the storms of our latitude pursue a definite course.[64]

In 1868, Cleveland Abbe, then director of the Cincinnati Observatory, revived the Smithsonian idea of meteorological forecasts, and suggested to the Cincinnati Chamber of Commerce that Cincinnati should be made the headquarters of meteorological observation for the United States, "for the purpose of collecting and comparing telegraphic weather reports from all parts of the land and making deductions therefrom." His proposals were favorably received, and he began, September 1, 1869, to issue the Weather Bulletin of the Cincinnati Observatory, which he continued until, in January, 1871, he was summoned to Washington to assist in organizing the national meteorological service, with which he has ever since been identified.

The Smithsonian meteorological system continued its functions until it was finally consigned to the custody of the Chief Signal Officer of the Army. Like all the efforts of this Institution, this work was in the direction of supplementing and harmonizing the work of all others, and attention was especially devoted to preparing and distributing blank forms in this direction, calculating and publishing extensive papers for systematizing observations, introducing standard instruments, collecting all public documents, printed matter, and manuscript records bearing on the meteorology of the American Continent, submitting these materials for scientific discussion, and publishing their results. The Smithsonian work was, during its whole existence, under the immediate personal direction of Professor Henry, assisted by Professor Arnold Guyot, who, in 1850,

prepared and published an exhaustive series of directions for meteorological observations, intended for the first-class observers cooperating
with the Smithsonian Institution.

The seeds planted by the army in 1819 began to bear perfect fruit fifty
years later, when, by act of Congress, in 1870, the Secretary of War was
authorized to carry into effect a scheme for "giving notice by telegraph
and signals of the approach and force of storms," and the organization of
a meteorological bureau adequate to the investigation of American
storms, and their preannouncement along the Northern lakes and the
seacoast was, under the auspices of the War Department, intrusted to
the Chief Signal Officer of the Army, Brigadier-General Albert J. Myer,
and a division, created in his office, was designated as the Division of
telegrams and reports for the benefit of commerce.

By a subsequent act of Congress, approved June 10, 1872, the Signal
Service was charged with the duty of providing such stations, signals, and
reports as might be found necessary for extending its research in the
interest of agriculture. In 1873, the work of the bureau of the division
having been eminently successful, and its successes having been recognized abroad as well as in this country, Congress, by a further act, authorized the establishment of signal-service stations at the light-houses and
life-saving stations on the lake seacoasts, and made provision for con-
necting them with telegraph lines or cables, "to be constructed, maintained, and worked under the direction of a chief signal officer of the
Army, or the Secretary of War and the Secretary of the Treasury," and in
this year also was begun the publication of a monthly Weather Review,
summarizing in a popular way all its data showing the result of its investigations, as well as presenting these in graphic weather charts.

In 1874 the entire system of Smithsonian weather observation in all
parts of the United States was transferred by Professor Henry to the
Signal Service. A few months previously, at the proposal of the Chief
Signal Officer, in the International Congress of Meteorologists convened

at Vienna, the system of world-wide cooperative simultaneous weather observations, since then so extensively developed, was inaugurated, and began to contribute its data to the Signal Office records. It is unnecessary to trace further the history of the beginning of the meteorological work of the Signal Service, but I doubt not that everyone at all familiar with its subsequent history, under the leadership of Generals Hazen and Greely, will agree with the opinion of Judge Daly, the president of the American Geographical Society, when he said that "nothing in the nature of scientific investigation by the National Government has proved so acceptable to the people, or has been so productive in so short a time of such important results, as the establishment of the Signal Service Bureau."[65]

The sixth President, John Quincy Adams, a man of culture broad and deep, found the presidency of the American Academy of Arts and Sciences so congenial to his tastes and sympathies that he did not hesitate to say that he prized it more highly than the chief magistracy of the nation. He considered his most important achievement to be the Report on Weights and Measures, prepared for Congress in 1818, and was justly proud of it, for it was a very admirable piece of scientific work, and is still considered the most important treatise on the subject ever written.

John Quincy Adams revived Washington's national university project, and made battle valiantly for an astronomical observatory.

In his first message to Congress afterwards, he said:

Among the first, perhaps the very first, instrument for the imprisonment of the condition of men is knowledge, and to the acquisition of much of the knowledge adapted to the wants, the comforts, and enjoyments of human life public institutions and seminaries of learning are essential. So convinced of this was the first of my predecessors in this office, now first in the memory, as, living, he was first in the hearts, of our country, that once and again in his addresses to the Congresses with whom he cooperated in the public service he earnestly recommended the establishment of seminaries of learning, to prepare for all the emergencies of peace and war—a national university and a military academy. With respect to the latter, had he lived to the present day, in turning his eyes to the institution at West Point he would have enjoyed the gratification of his most earnest wishes; but in surveying the city which has been honored with his name he would have seen the spot of earth which he had destined and bequeathed to the use and benefit of his country as the site for an university still bare and barren.[66]

And again:

Connected with the establishment of an university, or separate from it, might be undertaken the erection of an astronomical observatory, with provision for the support of an astronomer, to be in constant attendance of observation upon the phenomena of the heavens, and for the periodical publications of his observations. It is with no feeling of pride as an American that the remark may be made that on the comparatively small territorial surface of Europe there are existing upward of 130 of these light-houses of the skies, while throughout the whole American hemisphere there is not one. If we reflect a moment upon the discoveries which in the last four centuries have been made in the physical constitution of the universe by the means of these buildings and of observers stationed in them, shall we doubt of their usefulness to every nation? And while scarcely a year passes over our heads without bringing some new astronomical discovery to light, which we must fain receive at second hand from Europe, are we not cutting ourselves off from the means of returning light for light while we have neither observatory nor observer upon our half of the globe and the earth revolves in perpetual darkness to our unsearching eyes?

This appeal was received with shouts of ridicule; and the proposal "to establish a light-house in the skies" became a common byword which has scarcely yet ceased to be familiar. So strong was public feeling that, in the year 1832, in reviving an act for the continuance of the survey of the coast, Congress made a proviso, that "nothing in the act should be construed to authorize the construction or maintenance of a permanent astronomical observatory.[67]

Nothing daunted, Mr. Adams continued the struggle, and while a member of the House of Representatives, after his presidential term had expired, he battled for the observatory continually and furiously. An oration delivered by him in Cincinnati in 1843, closed with these words:

> Is there one tower erected to enable the keen-eyed observer of the heavenly vault to watch from night to night through the circling year the movement of the starry heavens and their unnumbered worlds? Look around you; look from the St. John to the Sabine; look from the mouth of the Neversink to the mouth of the Columbia, and you will find not one! or if one, not of our creation.

A correspondent of the London Athenæum, writing from Boston in May, 1840, spoke at length of the dearth of observatories in the United States, and of the efforts of John Quincy Adams to form a national astronomical establishment in connection with the Smithsonian bequest. The letter is of great interest as showing the state of opinion on scientific matters in America just half a century ago.

BOSTON, *May, 1840*

One of the prominent subjects of discussion among our *savans* . . . is the establishment of *Observatories* of a character suitable to our standing as a civilized nation, and still more to our exigencies as a practical, and especially as a commercial community. I verily believe that the yearly damage and destruction along our coast, for want of the securities which such institutions would supply, out-balances, beyond comparison, all it would cost to establish and maintain them in every principal city of the land. It is partly a sort of electioneering economy which leaves things thus, and which has heretofore refused or neglected to fit out Exploring Expeditions; to accumulate national treasures of art and science, and facilities for their prosecution; and generally to pursue a system of "in-breeding and cherishing," as Milton has it, "in a great people, the seeds of virtue and public civility;"—-excepting always what is done for the diffusion of elementary popular education. This education, to be sure, and this diffusion of it, we are taught to regard as necessaries in our moral and social being,—the "staff of public life" among us. And we are right. It is so. But there are many other things which we have *not* been taught to appreciate as they deserve, and the value of which we have gradually to grope our way to. Their day, however, will come; though it cannot be expected that either a government or a people, so youthful, so hurried, so fluctuating, can reach at once to the graces and the "fair humanities" of the old world. Remember that "The United States" are only some half-century old; and remember what we have been obliged to do and to suffer meanwhile, and under what circumstances. But, as I said before, the time is coming, if not come, when the heart of the nation shall acknowledge what is the high duty and destiny of a country like this; and *then*, I need not tell you, all is accomplished. Congress and the government must always represent the general, as well as the political character of the nation. It will be refined, scientific, public-spirited, or otherwise, as are the people. At this moment, as at all times, the representative and the represented, bear this relation to each other as intimately as might be expected from the nature of our institutions; and hence, from the signs which have appeared in the legislative bodies, I derive hope, and feel authorized to say what I have said of the advance, throughout our community, of what may be called the graceful and genial system of civilization, as distinguished from the practical and hard. This subject of observatories is quite in point. True, nothing has yet been done, but then a good deal has been said; and that is much: it is, in fact, *doing* much, in a case like this. It was something for Congress to bear being told what they had neglected, and patiently to discuss the subject.

The principal agent in bringing the subject forward has been Ex-President Adams, who, as you may be aware, is still an M. C., at the age of between seventy and eighty, and one of the halest and hardiest men in that body. His spirit is equal to his iron constitution. He spares himself no labour. So well is this understood, that it has been of late rather a practice to select the old gentleman for special burthens; and there are many matters of legislative action, which he really understands better, or knows better at least how to explore and determine, than any member of the House. Thus the Observatory business came upon him, at least indirectly; for, to some extent, he brought it on himself. You are, no doubt, familiar with the history of the great Smithsonian Bequest. When that business came before Congress, and especially as it was not a party one, all eyes were turned on Mr. Adams, and he was appointed Chairman of the Committee. In this capacity he has made sundry Reports: the last and ablest reviews the whole subject. In this he labours to show what general appropriation ought to be made of the fund—for that is not yet determined—and then to sustain a special recommendation, which is, to devote the income for about ten years to an *Observatory*, to be founded on national land, at Washington, "adapted to the most effective and continual observations of the phenomena of the heavens, and to be provided with the necessary, best, and most perfect instruments and books, for the periodical publication of the said observation, and for the annual composition and publication of a Nautical Almanack." The details of the plan may be omitted. Many, however, of the statistics connected with them, are new to us here, and of interest, including a Report on the British-establishments, furnished on request by the Astronomer Airy. To a greater extent these may be familiar to English readers, but perhaps not wholly so. I hope they do not know, for example, how much we deserve, as compared with other nations, the caustic strictures and lectures of Mr. Adams, who really gives us no quarter, being resolved not to spoil the child by sparing the rod, but rather to provoke us to find a remedy for the evils he describes. You yourself adverted, not long since, to the state of things among us, but only in general terms. The facts are these:—They have a small Observatory in process of erection at Tuscaloosa, Alabama, for the use of the University in that place. Professor Hopkins, of Williams' College, Massachusetts, has a little establishment of the short, and *this* is about all in that State,—all in New England! The only other establishment in the United States, known to me, is that in the Western Reserve College, Ohio, under the charge of Professor Loomis. Nothing of the kind at our national seat of government, or anywhere near it! Even Harvard University, "with all its antiquity, revenue, science, and renown," *has* thus far failed, though it appears that they are breaking ground at Cambridge; a house or houses having been purchased and fitted up, and one of our "savans" is already engaged in a series of magnetic and other observations. Now, how stands the case on your side the water? Why, in the British islands alone, there are observatories at the Universities of Cambridge and Oxford—at Edinburgh and Glasgow, in Scotland—and at Dublin and Armagh, in Ireland,—all receiving some patronage from the government—to say nothing of an observatory at the Cape of Good Hope; or of the establishments on the various remote and widely separated dependencies of the British Empire, including Van Diemen's Land, for the furnishing of which, we understand, arrangements have been made, in connexion with Captain Ross's expedition. In France, I believe, the provision is not less ample. On this part of the subject, Mr. Adams merely remarks, that the history of the Royal Observatory of that country would show the benefits conferred on mankind by the slightest notice bestowed by the rulers on the pursuit of knowledge; and that "the names of the four Cassinis would range in honourable distinction by the side of Flamsteed, Bradley, and Maskelyne."

Special reference is of course made to Greenwich, and Mr. Adams takes much pains to show how much that institution has done for science and for man. After recapitulating how by preserving observations we are indebted for a fixed standard for the measurement of *time*,—how, by the same science, man has acquired, so far as he possesses it, a standard for the measurement of *space*,—he observes, that the minutest of these observations contribute to the "increase and diffusion of knowledge" (the expressed object in Smithson's bequest). As to the more *brilliant*, we are reminded of an observation of Voltaire, that if the whole human race could be assembled from the creation of man to *his* time, in the gradation of genius, *Isaac Newton would stand at their head*; and the discoveries of Newton were the results of calculations, founded on the observations of others—of Copernicus, Tycho Brahe, Kepler, and Flamsteed. Greenwich has been considered rather an expensive establishment (among us), but Mr. Adams shows that, though costly, it has not been profitless.

Not to enter further into details of European countries, it appears that there are about one hundred and twenty *Observatories in continental Europe*; and that the most magnificent of them all has been lately founded by the Czar in the vicinity of his capital;—an enterprise sufficiently glorious, Mr. Adams observes, for the sovereign of such an empire; but the merit of which is vastly enhanced by the fact of its being undertaken and accomplished in such a latitude and climate:—"a region so near the pole, that it offers to the inspection of the human eye only a scanty portion of the northern hemisphere, with an atmosphere so chilled with cold and obscured with vapours, that it yields scarcely sixty days in the year when observation of the heavenly bodies is practicable." This last fact, it must be allowed, is rather an aggravation, or ought to be, to us republicans, *some* among whom affect to be special despisers of the bigoted Nicholas, and all his works. It seems, too, that *Mehemet Ali* has come forward as the patron of philosophical inquiry.

Thus matters stand at present, and Mr. Adams strongly urges prompt, practical action; and this scheme, with some modifications, and after our customary delays and discussions (in Congress) will be carried into execution, at least to a respectable extent. I am the more inclined to the opinion as it has been made clear in the progress of discussion that the establishments referred to *need* not be so enormously expensive as they generally are. In this matter we have been misled and discouraged by your own example, among others. We found that Cambridge Observatory cost £20,000, and that, among the instruments, the price of the mural circle alone was over £1,000, to say nothing of an equatorial telescope at £750, or a transit instrument at £600, and that as to Greenwich, the annual expenses, including salaries, repairs, and printing, exceeded £3,000. Now, this may be "sport for you," but it knocked our calculations on the head. Our ideas are not yet enlarged to that extreme point. To be sure, we can spend money for Florida wars; may, for better things—for internal improvements—for bridges over the Ohio river (St. Louis), or for market-houses and meeting houses of most liberal dimensions—for whatever, in a word, is practicable—as *we* understand it—and especially so much of it as private enterprise can execute without calling in government aid:—but ask for the adornments and muniments of art and science, in the ornamental or even in the scholar-like way, and it must be acknowledged the "sovereign people" move slow: they button their breeches' pockets and begin to "calculate." As to the Observatories, however, the case is better, for we find that much can be done at small expense. An establishment, of the merely *useful* kind, may be set up for a trifle. Not that Mr. Adams proposes to establish the National Observatory on such a scale. On the contrary, he thinks the Smithson Fund should be devoted to it for the present, and that not less than ten years of the income will be required. A more explicit estimate is also added, but it will be sufficient to observe that it comprises, besides a salary of $3,600 for the astronomer, funds for the compensation of four assistants, at $1,500 each, and two labourers, each at $600; for the purchase and procurement of instruments, $30,000; of which $20,000 might be applied for an assortment of the best instruments to be procured, and $10,000 for a fund, from the interest of which other instruments may be from time to time procured, and for repairs; for the library, $30,000, being $10,000 for first supply, and $20,000 for a fund for an income of $1,200 a year: and finally $30,000 for a fund, from the income of which, $1,800 a year, shall go to defray the expense of the yearly publication of the observations and of a Nautical Almanac.

It was the idea of Mr. Adams, in his later days, that the Smithson bequest, or at least its income for ten years, should be applied to the foundation of a national observatory and the publication of the Nautical Almanac, and he only abandoned it when an observatory had actually been established under the Navy Department in connection with the department of charts and instruments.

The establishment of an observatory had indeed been prominent in the minds of Washington and Jefferson, and was definitely proposed in Barlow's plan for a national institution, as well as in the project for a coast survey, submitted in 1837, in which it was proposed that there should be two observatories, formed at a fixed point, around which the survey, and particularly the nautical part of it, should be referred; their situation preferably to be in the State of Maine or lower Louisiana, since from them every celestial object observable, from the Tropics to the Arctic Circle and within about 20 degrees of longitude could be observed. Still, however, since various considerations might occasion the desire of placing one of these observatories in the city of Washington, just as observatories had been placed in the principal capitals of Europe, as a national object of scientific ornament, as well as a means for nourishing science in general, Hassler conceded that it might there be placed, since it would then be the proper place for the deposit of the standards of weights and measures, which also makes a special part the collection of instruments. James Monroe, when Secretary of State, in 1812, strongly urged upon Congress the establishment of an observatory, urging, first, the necessity of establishing a first meridian for the continent, and, in the second place, the fact that every enlightened nation had already established such an institution of learning. The immediate occasion for the intervention of the Secretary of State was the memorial of William Lambert, of Virginia, which was presented at various times from 1810 to 1821, and was accompanied by an elaborate report in 1822.

The action of Congress during the Adams Administration has been referred to. In 1830 Mr. Branch, of North Carolina, Secretary of the Navy under Jackson, strongly urged the establishment of an observatory for general astronomical purposes.

The beginning of the observatory seems to have been actually made on Capitol Hill during Mr. Adams's Administration, under instruction of Astronomers Lambert and Elliott, employed by Congress to determine the longitude of Washington. The President, in his diary of 1825, described a visit to Capitol Hill in company with Colonel Roberdeau, and spoke of witnessing an observation of the passage of the sun over the meridian, made with a small transit instrument. This instrument was very probably the one obtained by Hassler in Europe in 1815, which he never was permitted to use in connection with the Coast Survey work, and which passed into the hands of Lieutenant Wilkes in 1834, when it was placed in the small observatory, erected at his own expense, about a thousand feet north of the Dome of the Capitol.

It was at this establishment, which was known as the Naval Depot of Instruments, that the 5-foot transit was used, mainly for the purpose of reading the naval chronometer. When Wilkes went to sea with his expedition in 1837, Lieutenant James M. Gilliss became superintendent of the depot, and having obtained a 42-inch astronomical telescope, commenced a series of observations on the culmination of the moon and stars. In 1842 the establishment of a permanent depot of charts and instruments was authorized by Congress, and although the establishment of an observatory was not authorized in the bill, every effort was made by Lieutenant Gilliss and others interested in his work to secure suitable accommodations for astronomical work, and his plans having been approved by President Tyler, work was begun on the Naval Observatory, now known as the National Observatory.

There can be little doubt that the excellence of the work done by Gilliss himself, with his limited opportunities, did much to hasten the establishment of the observatory, and there is in this connection a traditional history. Encke's comet appeared in 1842, and was promptly observed by him. He read a paper concerning it before the National Institute. Senator Preston, an enthusiastic member of that organization, was present at the meeting. When Gilliss, still a very young man, shortly afterwards made a visit to the Senate committee room, the Senator remarked to him: "If you are the one who gave us notice of the comet, I will do all I can to help you."

A week afterwards a bill passed the Senate and House without formal discussion. The appropriation was $25,000, and although it was expressly for the establishment of a depot of charts and instruments, the report of the committee which had secured it was so emphatically in favor of astronomical, meteorological, and magnetic work that the Secretary of the Navy felt justified in assuming that Congress had sanctioned the broadest project for an observatory. Gilliss was at once sent abroad to obtain instruments and plans, while Lieutenant Matthew F. Maury was placed in charge of the depot, and when the observatory was completed in 1844 became its superintendent.

Maury's attitude toward astronomical work has been severely criticised, and, I think, misunderstood. He was, first of all, an enthusiastic officer of the Navy; second, an astronomer, and he deemed it appropriate that the chief effort of the office should be directed toward work which had a direct professional bearing. Although not neglecting astronomy (for under his direction two volumes of astronomical observations were published), his own attention, and oftentimes that of almost the entire office was devoted to hydrographic subjects. The work which he had accomplished was of the greatest practical importance to navigation, and nothing of a scientific nature up to that time accomplished in America received such universal attention and praise from abroad.

His personal popularity and his influence were very great, and the necessity for the maintenance of a national observatory was not in his day fully appreciated by the public. It is not at all impossible that, indirectly, through his meteorological and hydrographic work, he may have done more for the ultimate and permanent welfare of the National Observatory than could have been possible through exclusive attention to work of a purely astronomical character.

In 1861 Gilliss again became the Superintendent, and under his direction the Observatory took rank among the first in the world.

Before leaving the subject of the Observatory, reference should be made to astronomical work almost national in character accomplished in colonial days at Philadelphia under the direction of the American Philosophical Society, by which a committee of thirteen was appointed to make observations upon the transit of Venus in 1769.

Three temporary observatories were built, one in Philadelphia, one at Norristown, and one at Cape Henlopen. Instruments were imported from England, one of them a reflecting telescope with a Dollond micrometer, purchased in London by Doctor Franklin with money voted by the assembly of Pennsylvania. The transit was successfully observed and an elaborate report was published.

This enterprise is worthy of mention because it was the first serious astronomical work ever undertaken in this country. Being under the auspices of the only scientific society then in existence, it was in some sense a national effort. Had not the Revolution taken place, it would undoubtedly have resulted in the establishment of a well-equipped observatory in this country under the auspices of the home government. Doctor Thomas Ewing, the provost of the University of Pennsylvania, who seems to have been the first to propose the observations of 1769, and under whose direction they were carried on, visited London a few years later, and while there made interest with Lord North, the prime minister, and with Mr. Maskelyne, the astronomer royal, for the establishment of an observatory in Philadelphia, and that his efforts gave great promise of success may be shown by the letter here presented, addressed to him by Mr. Maskelyne in 1775:

GREENWICH, *August 4, 1775*

SIR: I received your late favor, together with your observations of the comet of 1770, and some [copies] of that of 1769, for which I thank you. I shall communicate [them] to the Royal Society, as you give me leave. In the present unhappy situation of American affairs, I have not the least idea that anything can be done toward erecting an observatory at Philadelphia, and therefore can not think it proper for me to take a part in any memorial you may think proper to lay before my Lord North at present. I do not mean, however, to discourage you from presenting a memorial from yourself. Were an observatory to be erected in that city, I do not know any person there more capable of taking care of it than yourself. Should Lord North do me the honor to ask my opinion about the utility of erecting an observatory at Philadelphia, I should then be enabled to speak out, being always a well-wisher to the promotion of science. You did not distinguish whether the times of your observations were apparent or mean time.

I am, your most humble servant,

N. MASKELYNE.

Rev. DR. EWING,

No. 25 Ludgate street.

In this connection mention should be made of the extended astronomical work done from 1763 to 1767, by Charles Mason, an assistant of Maskelyne, and Jeremiah Dixon, while surveying the boundary line between Pennsylvania and Maryland, and especially of the successful measurement by them of a meridian of latitude. Mason was a man of high scientific standing, but, though he became a citizen of Philadelphia, where he died in 1787, little is known of him beyond the record of his scientific work. He had been one of the observers of a transit of Venus at the Cape of Good Hope in 1761, and it was no doubt he who inspired the American Philosophical Society to its effort in 1769.

Another event in the Adams Administration was the beginning of the National Botanic Garden. The foundation of such an institution was one of the earliest of the projects for the improvement of the capital. Washington decided that it should be closely connected with the National University, on the site now occupied by the National Observatory, and stipulated that, should this site not be found available, another spot of ground, appropriated on the early maps to a marine hospital, might be substituted. The Columbian Institute, already referred to, had begun the formation of an arboretum as early as 1822, and in 1829 applied unsuccessfully to Congress for an appropriation to reimburse it for its expenditures. There was, however, no definite foundation until 1852, when the numerous living plants which had been brought back by the Wilkes Exploring Expedition in the Pacific, and which had for several years been kept in greenhouses adjoining the Patent Office, in which the natural-history collections of the expedition were kept, were removed to the present site of the Botanical Garden on the south side of Pennsylvania avenue just west of the Capitol. This garden was first under the direction of Mr. W. D. Brackenridge, who had been the horticulturist of the Wilkes Expedition. Mr. Brackenridge was succeeded by Mr. William R. Smith, a pupil of the Kew Botanical Garden, who has since been in charge of the establishment, and through whose industry it has been developed into a most creditable institution, which, it is hoped, may in time have an opportunity to exhibit its merits in a more suitable and less crowded locality.

Under Jackson, from 1829 to 1837, notwithstanding the remarkable commercial prosperity and an almost equal advance in literature, science did not prosper, and of actual progress there is little to record. The Coast Survey was reorganized under its original Superintendent, Hassler, in 1832, and Featherstonehaugh, an English geologist, made, in 1834, a re-connoissance in the elevated region between the Missouri and the Red River.

Van Buren's Administration, which began in 1837 and ended in 1841, presents more points of interest, for although the country was in a state of financial depression, his Cabinet was composed of extremely liberal and public-spirited men. Poinsett as Secretary of War, Kennedy as Secretary of the Navy, and other public men did much to promote science.

The United States Exploring Expedition was sent out under Captain Charles Wilkes, on a voyage of circumnavigation. Although published in an extremely limited edition, the magnificent volumes of its report are among the classics of scientific exploration.

The Wilkes Expedition was the first of the series of naval explorations which have contributed largely to science—Lynch's Dead Sea Expedition, Gilliss's Naval Astronomical Expedition to Chile, Herndon and Gibbons's Exploration of the Valley of the Amazons, Page's Paraguay Expedition, the Cruise of the Dolphin, Perry's Japan Expedition, Rogers's North Pacific Exploring Expedition, and the various expeditions made under the Hydrographic Office and the Coast Survey.

In 1840 two important national societies were founded—the National Institution for the Promotion of Science, and the American Society of Geologists and Naturalists—the one an association with a great membership, scientific and otherwise, including a large number of Government officials; the other composed exclusively of professional naturalists.

The purpose of each was the advancement of the scientific interests of the nation, which seemed more likely to receive substantial aid now that the money bequeathed by Smithson was lying in the Treasury vaults, waiting to be used.

The National Institution under the leadership of Joel R. Poinsett, of South Carolina, then Secretary of War, assisted by General J. J. Abert, F. A. Markoe, and others, had a short but brilliant career, which endured until the close of the Tyler Administration, and had an important influence on public opinion, bringing about in the minds of the people and of Congress a disposition to make proper use of the Smithson bequest, and which also did much to prepare the way for the National Museum. The extensive collections of the National Institution, and those of the Wilkes Expedition and other Government surveys were in time merged with those of the Smithsonian Institution, and having been greatly increased at the close of the Centennial Exposition, began in 1879 to receive substantial support from Congress.

The Society of Geologists was not so prominent at the time, but it has had a longer history, for in 1850 it became the American Association for the Advancement of Science. Although it dated its origin from 1840, it was essentially a revival and continuation of the old American Geological Society, organized September 6, 1819, in the philosophical room of Yale College, and in its day a most important body. Its members, following European usage, appended to their names the symbols M. A. G. S., and among them were many distinguished men, for at that time almost every one who studied any other branch of science cultivated geology also.

The American Association prepared the way for the National Academy of Sciences, which was established by Congress in 1863, having for its first president Alexander Dallas Bache, who in his presidential address at the second meeting of the American Association, twelve years before, had pointed out the fact that "an institution of science supplementary to existing ones is much needed to guide public action in reference to scientific matters,"[68] and whose personal influence was very potent in bringing that institution into existence. In advocating before Congress the plan for the National Academy of Sciences, Senator Sumner avowedly followed the lead of Joel Barlow, the projector of the National Institution in 1806.[69]

The system of national scientific organizations thus inaugurated is still expanding. Within the past few years there have sprung into existence a considerable number of learned societies devoted to special subjects, usually with unlocalized membership, and holding meetings from year to year in different cities. Among these are those named below:

The American Anatomical Society
The American Dialect Society
The American Folk-lore Society
The American Geographical Society (of New York) and the National Geographic Society (of Washington)
The American Geological Society
The American Historical Association
The American Institute of Mining Engineers
The American Meteorological Society
The American Metrological Society

The American Oriental Society
The American Ornithologists' Union
The American Philological Association
The American Physiological Society
The American Society of Naturalists
The American Society for Psychical Research
The Archæological Institute of America
The Botanical Club of the American Association
The Franklin Institute

That the organization of such societies has been so long delayed was perhaps due to the fact that during the first six decades of the century the number of scientific investigators was comparatively small, and scientific work of original character was confined to a few of the large cities, so that local organizations, supplemented by the annual summer meetings of the American Association for the Advancement of Science, answered all needs. Since the close of the civil war, and of the period of ten years which elapsed before our country was restored to commercial prosperity, and indeed before it had begun to fully feel the effects of the great scientific renaissance which originated in 1859 with the publication of Darwin's Origin of Species, there has been a great increase in the number of persons whose time is chiefly devoted to original scientific work.

Nothing has contributed so materially to this state of affairs as the passage by Congress in 1862 of the bill, introduced by the Hon. Justin S. Morrill, of Vermont, to establish scientific and industrial educational institutions in every State, supplemented in 1887 by the Hatch bill for the founding of the agricultural experiment stations.[70] The movement was at first unpopular among American educators, but after a quarter of a century of trial the land-grant college system has not only demonstrated its right to exist, but is by many regarded as forming one of the chief strongholds of our national scientific prosperity.[71]

One of the most important effects of the movement has been to stimulate the establishment of State scientific schools and universities, and every one of the forty-two Commonwealths has already a university or a college performing, or intended to perform, university functions.

It is worthy of remark that with six exceptions every State has in less than twenty years of its admission had a State college or university of its own. Only twelve have delayed more than ten years, and fifteen have come into the Union already equipped. Ten of these were colonies and original States. All but one of the remainder were those admitted in 1889, for each of our four new States was provided with the nucleus of a State university before it sought admission to the Union. Twenty-eight of the State and Territorial universities had their origin in land grants from the General Government other than those for agricultural and mechanical colleges.[72]

The completeness of the State system of scientific educational institutions is in marked contrast with that of the scientific societies in the same States, organized by the direct action of the people rather than by government.

Academies of science bearing the names of the States of our confederation and often sanctioned by their laws, may be regarded as in some sense national. Although nearly all of our States have historical societies, only twelve of the forty-two have academies of science, or organizations which are their equivalent. That there should be in 1889 thirty States without academies of science, and fourteen States and Territories in which there are no scientific societies of any description whatever, is a noteworthy fact.[73]

During Van Buren's Presidency, the Department of Agriculture had its formal beginning.

The chief promoter of this idea was Henry L. Ellsworth, of Connecticut, Commissioner of Patents, whose efforts culminated twenty-six years later in the establishment of a department, and, after another period of twenty-six years, in the elevation of the head of that Department to the dignity of a Cabinet officer. Ellsworth began work by distributing seeds and plants for experimental culture, acquiring these without expense, and sending them out under the franks of friendly Congressmen. After three years (in 1839) Congress recognized the value of the work in this direction by appropriating $1,000 from the Patent Office fund to enable him to collect and distribute seeds, to collect agricultural statistics, and to make agricultural investigations. Appointed by Jackson in 1836, Ellsworth served through the two successive terms of Van Buren and Tyler, and in his nine years of official work his devotion to the interests of agriculture produced excellent results, and placed the service on a firm foundation. Though Newton was in name the first Commissioner of Agriculture, Ellsworth deserves to be kept in memory as the real founder of the Department.

The appropriations at first were insignificant, and occasionally, as in 1841, 1842, and 1846, Congress seems to have forgotten to make any provision whatever for the work, which consequently went forward under difficulties. In 1853 the first appropriation directly for agriculture was made, in 1855 the whole amount up to that time withdrawn for this purpose from the Patent Office fund was reimbursed, and from that time on the money grants became yearly larger, and the work was allowed slowly to expand. The seed work increased, and in 1856 a propagating garden was begun. The Agricultural Report, which began in 1841, and was until 1862 printed as a part of that of the Patent Office, became yearly more extensive, and showed a general average annual growth in value. In 1854 work in economic entomology began, with the appointment of Townend Glover to investigate and report upon the habits of insects injurious and beneficial to agriculture. In 1855 the chemical and botanical divisions were inaugurated.

David P. Holloway, of Indiana, the thirteenth Commissioner of Patents, was instrumental in effecting a most important reform in the scientific administration of the Government. In his first annual report, made in January, 1862, he advocated enthusiastically the creation of a department of the productive arts, to be charged with the care of agriculture and all the other industrial interests of the country, and he was so far successful that on May 15 Congress established the Department of Agriculture. The first Commissioner was Isaac Newton, who had been for a year or more superintendent of the agricultural division of the Patent Office. From 1862 to 1889 there were six Commissioners: Newton (1862–1867), Capron (1867–1871), Watts (1871–1877), Le Duc (1877–1881), Loring

(1881–1885), and Colman (1885–1889), and under the administration of each important advances were made, and the value of the work became yearly greater. Buildings were erected, a chemical laboratory established, the departments of animal industry, economic ornithology and mammalogy, pomology, vegetable pathology, silk culture, microscopic, forestry, and experiment stations were added, and the system of publications greatly extended. The Department, as now organized, is one of the most vigorous of our national scientific institutions, and with its powerful staff and close affiliations with the forty-six State agricultural experiment stations, manned as they are by nearly four hundred trained investigators, it has possibilities for the future which can scarcely be overestimated.[74]

The term of the ninth President was too short to afford matter for comment. It should be mentioned, however, that General Harrison published in Cincinnati, in 1838, A Discourse on the Aborigines of the Valley of the Ohio, and was the only President, except Jefferson and John Quincy Adams, who has ever produced a treatise upon a scientific theme.

In 1841 John Tyler, of Virginia, became President. His period of administration was a stormy one, and the atmosphere of Washington at that time was not favorable for scientific progress. During this Administration, however, important reforms took place in the organization of the Navy, which resulted in great benefit to science. These were largely the result of the interest of Hon. A. P. Upshur, Secretary of the Navy, at whose instance President Tyler abolished the existing Board of Naval Commissioners and vested the authority formerly exercised by them in separate bureaus. To many of the pressing necessities for reform of the service, Lieutenant Maury had called attention in his essays, published in the Southern Literary Messenger, under the title of Scraps from a Lucky Bag, and over the signature of Harry Bluff. As a result of this movement, experiments in applying steam to war vessels were actively prosecuted, and the first bill was passed for the establishment at Annapolis of the United States Naval Academy, finally accomplished in 1845, and a little later (in 1848) the position of the professors of mathematics in the Navy was dignified and improved, and their numbers limited, with manifest advantage to the scientific service of the Government.[75]

Indirectly, the reorganization of the Navy had a powerful influence in the development of the Coast Survey, which was reorganized in 1843–44, with Alexander Dallas Bache as its superintendent, for this new system afforded ample means to that organization for ascertaining the topography of this coast and making contributions to the science of ocean physics.

Another enterprise was the sending of the Frémont exploring expedition to California and Oregon. It is interesting to know that Captain Frémont was appointed the leader of this expedition against the indignant protests of the topographical engineers, who insisted that a graduate of West Point should be chosen.[76]

The final establishment of the Naval Observatory took place also at this time. The history of this enterprise from the scientific standpoint, has already been discussed, but it may be well to note that it derived its chief political support from Mr. Upshur, then Secretary of the Navy.[77]

To this period belongs also the promotion of experiments with the electric telegraph by our Government. The line from Washington to Baltimore was erected by means of an appropriation of $30,000, the passage of which was warmly urged by the President, who fifteen years later wrote the following letter, full of historical reminiscences:

SHERWOOD FOREST, *September 1, 1858*
To his Honor the mayor, and to the Honorable the Common Council of the City of New York:

GENTLEMEN: In consequence of my absence from this place, I did not receive until to-day your polite invitation to be present at the festivities of to-day, and the municipal dinner to be given to Cyrus W. Field, Esq., and others at the Metropolitan Hotel to-morrow, in commemoration of the laying of the "Atlantic Cable." To be present, therefore, at the time appointed is a thing impossible. All that I can do is to express my cordial concurrence with you in according all praise to those through whose indomitable energy this great work has been accomplished.

When, in 1843, a modest and retired gentleman, the favored child of science, called upon me at the executive mansion, to obtain from me some assurance of my cooperation with him in procuring from Congress a small appropriation to enable him to test his great invention; and when at an after-day I had the satisfaction of placing my signature in approval of the act making an appropriation of $30,000 to enable him to connect Washington with Baltimore by his telegraph wire; and when at a still later day I had the pleasure, from the basement of the Capitol, to exchange greetings with the Chief-Justice of the United States, who was at the Baltimore end of the line, I confess that it had not entered my mind that not only was lightning to become the messenger of thought over continents of dry lands, but that the same all pervading agent was to descend into the depths of the ocean, far below the habitations of living things, and over those fathomless depths to convey, almost in the twinkling of an eye, tidings from nation to nation, and continent to continent. To the great inventor of this, the greatest invention, is due the laurel wreath that can never wither, and to those that have given it a habitation and a home in the waters of the great deep all praise is due.

With sentiments of high consideration, I have the honor to be, most respectfully and truly yours, etc.,

JOHN TYLER.

President Polk served from 1845 to 1849. During this period was organized the Smithsonian Institution, which, though it bears the name of a private citizen and a foreigner, has been for nearly half a century one of the principal rallying points of the scientific workers of America. It has also been intimately connected with very many of the most important scientific undertakings of the Government.

Many wise and enlightened scholars have given to the Smithsonian

Institution the best years of their lives, and some of the most eminent scientific men of our country have passed their entire lifetime in work for its success. Its publications, six hundred and seventy in number, which when combined make up over one hundred dignified volumes, are to be found in every important library in the world, and some of them, it is safe to say, on the working table of every scientific investigator in the world who can read English.

Through these books, through the reputation of the men who have worked for it and through it, and through the good accomplished by its system of international exchange, by means of which within the past thirty-eight years 1,262,114 packages of books and other scientific and literary materials have been distributed to every region of the earth, it has acquired a reputation at least as far-reaching as that of any other institution of learning in the world.

No one has been able to show why Smithson selected the United States as the seat of his foundation. He had no acquaintances in America, nor does he appear to have had any books relating to America except two. Rhees quotes from one of these, Travels through North America, by Isaac Weld, secretary of the Royal Society, a paragraph concerning Washington, then a small town of five thousand inhabitants, in which it is predicted that "the Federal city, as soon as navigation is perfected, will increase most rapidly, and that at a future day, if the affairs of the United States go on as prosperously as they have done, it will become the grand emporium of the West, and rival in magnitude and splendor the cities of the old world."

Inspired by a belief in the future greatness of the new nation, realizing that while the needs of England were well met by existing organizations such as would not be likely to spring up for many years in a new, poor, and growing country, he founded in the new England an institution of learning, the civilizing power of which has been of incalculable value.

Who can attempt to say what the condition of the United States would have been to-day without this bequest?

In the words of John Quincy Adams:

Of all the foundations of establishments for pious or charitable uses which ever signalized the spirit of the age, or the comprehensive beneficence of the founder, none can be named more deserving the approbation of mankind.

The most important service by far which the Smithsonian Institution has rendered to the nation from year to year since 1846—intangible, but none the less appreciable—has been its constant cooperation with the Government, public institutions, and individuals in every enterprise, scientific or educational, which needed its advice, support, or aid from its resources.

There have been, however, material results of its activities, the extent of which can not fail to impress anyone who will look at them; the most important of these are the library and the museum, which have grown up under its fostering care.

The library has been accumulated without aid from the Treasury of the United States; it has, in fact, been the result of an extensive system of exchanges, the publications of the institution having been used to obtain similar publications from institutions of learning in all parts of the world.

In return for its own publications the Institution has received the great collection of books which form its library.

This library, consisting of more than a quarter of a million volumes and parts of volumes, has for over twenty years been deposited at the Capitol as a portion of the Congressional Library, and is constantly being increased. In the last fiscal year nineteen thousand titles were thus added to the national collection of books.

Chiefly through its exchange system the Smithsonian had, in 1865, accumulated about forty thousand volumes, largely publications of learned societies, containing the record of the actual progress of the world in all that pertains to the mental and physical development of the human family, and affording the means of tracing the history of at least every branch of positive science since the days of revival of letters until the present time.

The books, in many cases presents from old European libraries, and not to be obtained by purchase, formed even then one of the best collections of the kind in the world.

The danger incurred from the fire of that year, and the fact that the greater portion of these volumes, being unbound and crowded into insufficient space, could not be readily consulted, while the expense to be incurred for this binding, enlarged room, and other purposes connected with their use threatened to grow beyond the means of the Institution, appear to have been the moving causes which determined the regents to accept an arrangement by which Congress was to place the Smithsonian library with its own in the Capitol, subject to the right of the Regents to withdraw the books on paying the charges of binding, etc. Owing to the same causes (which have affected the Library of Congress itself) these principal conditions, except as regards their custody in a fireproof building, have never been fulfilled.

The books are still deposited chiefly in the Capitol, but though they have now increased from 40,000 to fully 250,000 volumes and parts of volumes, forming one of the most valuable collections of the kind in existence, they not only remain unbound, but in a far more crowded and inaccessible condition than they were before the transfer. It is hardly necessary to add that these facts are deplored by no one more than by the Librarian of Congress.

The purchasing power of the publications of the Institution, when offered in exchange, is far greater than that of money, and its benefit is exerted chiefly in behalf of the National Library, and also, to a considerable extent, in behalf of the National Museum.

The amount expended during the past forty years from the private fund of the Institution in the publication of books for gratuitous distribution has been $350,000, a sum nearly half as great as the original Smithson bequest.

These publications have had their influence for good in many ways, but, in addition to this, a library much more than equal in value to the outlay has, through their buying power, come into the possession of the nation.

In addition to all this, a large amount of material has been acquired for the Museum by direct expenditure from the private fund of the Smithsonian Institution. The value of the collections thus acquired is estimated to be more than equal to the whole amount of the Smithsonian bequest.

The early history of the Museum was much like that of the library. It was not until 1858 that it became the authorized depository of the scientific collections of the Government, and it was not until after 1876 that it was officially recognized as the National Museum of the United States.

But for the provident forethought of the organizers of the Smithsonian Institution the United States would probably still be without even a reputable nucleus for a national museum or a scientific library.

For nearly half a century the Institution has been the object of the watchful care of many of America's most enlightened public men. Vice-Presidents Fillmore and Dallas, and Roger B. Taney, Salmon P. Chase, Morrison R. Waite, and Melville W. Fuller, Chief Justices of the United States, have in succession occupied the Chancellor's chair. George Bancroft, John C. Breckinridge, Lewis Cass, Rufus Choate, Samuel S. Cox, Schuyler Colfax, Garrett Davis, Jefferson Davis, Stephen A. Douglas,

William H. English, William P. Fessenden, James A. Garfield, Hannibal Hamlin, Henry W. Hilliard, George P. Marsh, James M. Mason, Justin S. Morrill, Robert Dale Owen, James A. Pearce, William C. Preston, Richard Rush, General W. T. Sherman, Lyman Trumbull, and William A. Wheeler have been at various times leaders in the deliberations of the Board of Regents.

The representatives of science on the Board, Professor Agassiz, Professor Bache, Professor Coppee, Professor Dana, General Delafield, Professor Felton, Professor Gray, Professor McLean, General Meigs, President Porter, General Totten, and Dr. Welling, have usually held office for long periods of years, and have given to its affairs the most careful attention and thought.

The relation of the Smithsonian Institution to the Government has been unique and unparalleled elsewhere. No one will question the assertion that the results of its work have been far wider than those which its annual reports have ever attempted to show forth.

During the administration of Van Buren and the succeeding ones, governmental science, stimulated by Bache, Henry, and Maury, scientific administrators of a new and more vigorous type than had been previously known in Washington, rapidly advanced, and prior to 1861 the institutions then existing had made material progress.

Those of more recent growth, such as the Army Medical Museum, founded in 1862;[78] the Bureau of Education, founded in 1867;[79] the Fish Commission, founded in 1870;[80] the Bureau of Ethnology, founded in 1879,[81] although not less important than many of those already discussed, are so recent in origin that the events connected with their development have not passed into the domain of history.

The material results of the scientific work of the Government during the past ten years undoubtedly surpass in extent all that had been accomplished during the previous hundred years of the independent existence of the nation. With this recent period the present paper has no concern, for it has been written from the standpoint of Carlyle, who, in Sartor Resartus, states his belief that "in every phenomenon the beginning remains always the most notable moment."

It is nevertheless very encouraging to be assured that the attitude of our Government toward scientific and educational enterprises is every year becoming more and more in harmony with the hopes of the founders of our Republic, and in accord with the views of such men as Washington, Franklin, Jefferson, John Adams, Madison, Monroe, John Quincy Adams, Gallatin, and Rush.

It is also encouraging to know that the national attitude toward science is the subject of constant approving comment in Europe. Perhaps the most significant recent utterance was that of Sir Lyon Playfair in his address before the British Association for the Advancement of Science, at the Aberdeen meeting. He said:

On September 14, 1859, I sat on this platform and listened to the eloquent address and wise counsel of the Prince Consort. At one time a member of his household, it was my privilege to cooperate with this illustrious prince in many questions relating to the advancement of science. I naturally, therefore, turned to his presidential address to see whether I might not now continue those counsels which he then gave with all the breadth and comprehensiveness of his masterly speeches. I found, as I expected, a text for my own discourse in some pregnant remarks which he made upon the relation of Science to the State. They are as follows: "We may be justified in hoping . . . that the Legislature and the State will more and more recognize the claims of science to their attention, so that it may no longer require the begging-box, but speak to the State like a favored child to its parent, sure of his paternal solicitude for its welfare; that the State will recognize in science one of its elements of strength and prosperity, to foster which the clearest dictates of self-interest demand."

This opinion, in its broadest sense, means that the relations of science to the State should be made more intimate because the advance of science is needful to the public weal.

The importance of promoting science as a duty of statecraft was well enough known to the ancients, especially to the Greeks and Arabs, but it ceased to be recognized in the dark ages, and was lost to sight during the revival of letters in the fifteenth and sixteenth centuries. Germany and France, which are now in such active competition in promoting science, have only publicly acknowledged its national importance in recent times. Even in the last century, though France had its Lavoisier and Germany its Leibnitz, their Governments did not know the value of science. When the former was condemned to death in the Reign of Terror, a petition was presented to the rulers that his life might be spared for a few weeks in order that he might complete some important experiments, but the reply was, "The Republic has no need of savants." Earlier in the century the much-praised Frederick William of Prussia shouted with a loud voice, during a graduation ceremony in the University of Frankfort, "An ounce of mother-wit is worth a ton of university wisdom." Both France and Germany are now ashamed of these utterances of their rulers, and make energetic efforts to advance science with the aid of their national resources. More remarkable is it to see a young nation like the United States reserving large tracts of its national lands for the promotion of scientific education. In some respects this young country is in advance of all European nations in joining science to its administrative offices. Its scientific publications . . . . are an example to other Governments. The Minister of Agriculture is surrounded with a staff of botanists and chemists. The Home Secretary is aided by a special Scientific Commission to investigate the habits, migrations, and food of fishes, and the latter has at his disposal two specially-constructed steamers of large tonnage.

In the United Kingdom we are just beginning to understand the wisdom of Washington's farewell address to his countrymen, when he said: "Promote as an object of primary importance institutions for the general diffusion of knowledge. In proportion as the structure of government gives force to public opinion, it is essential that public opinion should be enlightened.

APPENDIX A
PLAN OF A FEDERAL UNIVERSITY

[*From the Pennsylvania Gazette, 1788. Quoted in the Massachusetts Centinel,
Saturday, November 29, 1788.*]

"Your government cannot be executed. It is too extensive for a republick. It is
contrary to the habits of the people," say the enemies of the Constitution of
the United States. However opposite to the opinions and wishes of a majority
of the citizens of the United States these declarations and predictions may be,
they will certainly come to pass, unless the people are prepared for our new
form of government, by an education adapted to the new and peculiar situation
of our country.—To effect this great and necessary work, let one of the first
acts of the new Congress be, to establish within the district to be allotted for
them, a FEDERAL UNIVERSITY into which the youth of the United States shall be
received after they have finished their studies, and taken degrees in the colleges
of their respective States. In this University let those branches of literature only
be taught, which are calculated to prepare our youth for civil and publick life.
These branches should be taught by means of lectures, and the following arts
and sciences should be the subject of them:

1. The principles and forms of government, applied in a particular manner
to the explanation of every part of the constitution and laws of the United
States, together with the laws of nature and nations, which last should include
everything that relates to peace, war, treaties, ambassadors, and the like.

2. History, both ancient and modern, and chronology.

3. Agriculture in all its numerous and extensive branches.

4. The principles and practice of manufactures.

5. The history, principles, objects, and channels of commerce.

6. Those parts of mathematicks which are necessary to the division of property, to finance, and to the principles and practice of war—for there is too
much reason to fear that war will continue, for some time to come, to be the
unchristian mode of deciding disputes between christian nations.

7. Those parts of natural philosophy and chemistry, which admit of an application to agriculture, manufacture, commerce, and war.

8. Natural history, which includes the history of animals, vegetables, and fossils. To render instruction in these branches of science easy, it will be necessary to establish a museum, as also a garden, in which not only all the shrubs, etc., but all the forest trees of the United States should be cultivated. The great Linnæus, of Upsal enlarged the commerce of Sweden, by his discoveries in natural history. He once saved the Swedish navy by finding out the time in which a worm laid its eggs, and recommending the immersion of the timber, of which the ships were built, at that season wholly under water. So great were the services this illustrious naturalist rendered his country by the application of his knowledge to agriculture, manufactures, and commerce, that the present King of Sweden pronounced an eulogium upon him from his throne, soon after his death.

9. Philology, which should include, besides rhetorick and criticism, lectures upon the construction and pronunciation of the English language. Instruction in this branch of literature will become the more necessary in America, as our intercourse must soon cease with the bar, the stage, and the pulpits of Great Britain, from whence we receive our knowledge of the pronunciation of the English language. Even modern English books should cease to be the models of stile in the United States. The present is the age of simplicity in writing in America. The turgid stile of Johnson—the purple glare of Gibbon, and even the studied and thickset metaphours of Junius, are all equally unnatural, and should not be admitted into our country. The cultivation and perfection of our language becomes a matter of consequence when viewed in another light. It will probably be spoken by more people in the course of two or three centuries, than ever spoke any one language at one time since the creation of the world. When we consider the influence which the prevalence of only *two* languages, viz, the English and the Spanish, in the extensive regions of North and South America, will have upon manners, commerce, knowledge, and civilization, scenes of human happiness and glory open before us, which elude from their magnitude the utmost grasp of the human understanding.

10. The German and French languages should be taught in this University.

The many excellent books which are written in both these languages upon all subjects, more especially upon those which relate to the advancement of national improvements of all kinds, will render a knowledge of them an essential part of the education of a legislator of the United States.

11. All those athletick and manly exercises should likewise be taught in the University, which are calculated to impart health, strength, and elegance to the human body.

To render the instruction of our youth as easy and extensive as possible in several of the above-mentioned branches of literature, let four young men of good education and active minds be sent abroad at the publick expense, to collect and transmit to the professors of the said branches all the improvements that are daily made in Europe, in agriculture, manufactures, and commerce, and in the arts of war and practical government. This measure is rendered the more necessary from the distance of the United States from Europe, by which means the rays of knowledge strike the United States so partially, that they can be brought to a useful focus, only by employing suitable persons to collect and transmit them to our country. It is in this manner that the northern nations of Europe have imported so much knowledge from their southern neighbors, that the history of the agriculture, manufactures, commerce, revenues, and military of *one* of these nations will soon be alike applicable to all of them.

Besides sending four young men abroad to collect and transmit knowledge for the benefit of our country, *two* young men of suitable capacities should be employed at the publick expense in exploring the vegetable, mineral, and animal productions of our country, in procuring histories and samples of each of them, and in transmitting them to the professor of natural history. It is in consequence of the discoveries made by young gentlemen employed for these purposes, that Sweden, Denmark, and Russia have extended their manufactures and commerce, so as to rival in both the oldest nations in Europe.

Let the Congress allow a liberal salary to the Principal of this University. Let it be his business to govern the students, and to inspire them by his conversation, and by occasional publick discourses, with federal and patriotick sentiments. Let this Principal be a man of extensive education, liberal manners, and dignified deportment.

Let the Professors of each of the branches that have been mentioned, have a moderate salary of 150 or 200 pounds a year, and let them depend upon the number of their pupils to supply the deficiency of their maintenance from their salaries. Let each pupil pay for each course of lectures two or three guineas.

Let the degrees conferred in this University receive a new name, that shall designate the design of an education for civil and publick life. Should this plan of a federal University, or one like it be adopted, then will begin the golden age of the United States. While the business of education in Europe, consists in lectures upon the ruins of Palmyra and the antiquities of Herculaneum; or in dispute about Hebrew points, Greek particles, or the accent and quantity of the Roman language, the youth of America will be employed in acquiring those branches of knowledge which increase the convenience of life, lessen human misery, improve our country, promote population, exalt the human understanding, and establish domestick, social, and political happiness.

Let it not be said, "that this is not the *time* for such a literary and political establishment. Let us first restore publick credit, by funding or paying our debts—let us regulate our militia—let us build a navy—and let us protect and extend our commerce. After this, we shall have leisure and money to establish a University for the purposes that have been mentioned." This is false reasoning. We shall never restore publick credit—regulate our militia—build a navy—or revive our commerce, until we remove the ignorance and prejudices, and change the habits of our citizens, and this can never be done until we inspire them with federal principles, which can only be effected by our young men meeting and spending two or three years together in a national University, and afterwards disseminating their knowledge and principles through every county, town, and village of the United States. Until this is done—Senators and Representatives of the United States, you will undertake to make bricks without straw. Your supposed union in Congress will be a rope of sand. The inhabitants of Massachusetts began the business of government by establishing the University of Cambridge, and the wisest kings in Europe, have always found their literary institutions the surest means of establishing their power, as well as of promoting the prosperity of their people.

These hints for establishing the Constitution and happiness of the United

States upon a permanent foundation, are submitted to the friends of the federal government, in each of the States, by a private CITIZEN OF PENNSYLANIA. [*sic*]

APPENDIX B
ADDRESS TO THE PEOPLE OF THE UNITED STATES, BY BENJAMIN RUSH, M.D., 1787

[*Reprinted from Niles's Principles and Acts of the Revolution in America, pp. 234–236.*]

There is nothing more common, than to confound the terms of *American Revolution* with those of *the late American war*. The American war is over: but this is far from being the case with the American revolution. On the contrary, nothing but the first act of the great drama is closed. It remains yet to establish and perfect our new forms of government; and to prepare the principles, morals, and manners of our citizens, for these forms of government, after they are established and brought to perfection.

The confederation, together with most of our state constitutions, were formed under very unfavorable circumstances. We had just emerged from a corrupted monarchy. Although we understood perfectly the principles of liberty, yet most of us were ignorant of the forms and combinations of power in republics. Add to this, the British army was in the heart of our country, spreading desolation wherever it went: our resentments, of course, were awakened. We deserted the British name, and unfortunately refused to copy some things in the administration of justice and power, in the British government, which have made it the admiration and envy of the world. In our opposition to monarchy, we forgot that the temple of tyranny has two doors. We bolted one of them by proper restraints; but we left the other open, by neglecting to guard against the effects of our own ignorance and licentiousness. Most of the present difficulties of this country arise from the weakness and other defects of our governments.

My business at present shall be only to suggest the defects of the confederation. These consist—1st. In the deficiency of coercive power. 2d. In a defect of exclusive power to issue paper money and regulate commerce. 3d. In vesting the sovereign power of the United States in a single legislature: and, 4th. In the too frequent rotation of its members.

A convention is to sit soon for the purpose of devising means of obviating part of the two first defects that have been mentioned. But I wish they may add to their recommendations to each state, to surrender up to congress their power of emitting money. In this way, a uniform currency will be produced, that will facilitate trade, and help to bind the states together. Nor will the states be deprived of large sums of money by this means, when sudden emergencies require it; for they may always borrow them, as they did during the war, out of the treasury of congress. Even a loan office may be better instituted in this way, in each state than in any other.

The two last defects that have been mentioned, are not of less magnitude than the first. Indeed, the single legislature of congress will become more dangerous, from an increase of power, than ever. To remedy this, let the supreme federal power be divided, like the legislatures of most of our states, into two distinct, independent branches. Let one of them be styled the council of the states and the other the assembly of the states. Let the first consist of a single delegate—and the second, of two, three, or four delegates, chosen annually by each state. Let the president be chosen annually by the joint ballot of both houses; and let him possess certain powers, in conjunction with a privy council, especially the power of appointing most of the officers of the United States. The officers will not only be better, when appointed this way, but one of the principal causes of faction will be thereby removed from congress. I apprehend this division of the power of congress will become more necessary, as soon as they are invested with more ample powers of levying and expending public money.

The custom of turning men out of power or office, as soon as they are qualified for it, has been found to be absurd in practice. Is it virtuous to dismiss a general—a physician—or even a domestic, as soon as they have acquired knowledge sufficient to be useful to us, for the sake of increasing the number of able generals, skillful physicians—and faithful servants? We do not. Government is a science, and can never be perfect in America, until we encourage men to devote not only three years, but their whole lives to it. I believe the principal reason why so many men of abilities object to serving in congress, is owing to their not thinking it worth while to spend three years in acquiring a profession, which their country immediately afterwards forbids them to follow.

There are two errors or prejudices on the subject of government in America, which lead to the most dangerous consequences.

It is often said, "that the sovereign and all other power is seated *in* the people." This idea is unhappily expressed. It should be—"all power is derived *from* the people," they possess it only on the days of their elections. After this, it is the property of their rulers; nor can they exercise or resume it, unless it be abused. It is of importance to circulate this idea, as it leads to order and good government.

The people of America have mistaken the meaning of the word sovereignty: hence each state pretends to be *sovereign*. In Europe, it is applied only to those states which possess the power of making war and peace—of forming treaties, and the like. As this power belongs only to congress, they are the only *sovereign* power in the United States.

We commit a similar mistake in our ideas of the word independent. No individual state, as such, has any claim to independence. She is independent only in a union with her sister states in congress.

To conform the principles, morals, and manners of our citizens, to our republican forms of government, it is absolutely necessary, that knowledge of every kind should be disseminated through every part of the United States.

For this purpose, let congress, instead of laying out a half million of dollars, in building a federal town, appropriate only a fourth of that sum, in founding a federal university. In this university let everything connected with government, such as history—the law of nature and nations—the civil war—the municipal laws of our country—and the principles of commerce—be taught by competent professors. Let masters be employed, likewise, to teach gunnery—fortification—and everything connected with defensive and offensive war. Above all, let a professor of, what is called in the European universities, economy, be established in this federal seminary. His business should be to unfold the principles and practice of agriculture and manufactures of all kind, and to enable him to make his lectures more extensively useful, congress should support a traveling correspondent for him, who should visit all the nations of Europe, and transmit to him, from time to time, all the discoveries and improvements that are made in agriculture and manufactures. To this seminary, young men should be encouraged to repair, after completing their academical studies in the colleges of their respective states. The honors and offices of the United States should, after a while, be confined to persons who had imbibed federal and republican ideas in this university.

For the purpose of diffusing knowledge, as well as extending the living principle of government to every part of the United States—every state—city—county—village—and township in the union should be tied together by means of the post-office. This is the true nonelectric wire of government. It is the only means of conveying heat and light to every individual in the federal commonwealth. "Sweden lost her liberties," says the abbe Raynal, "because her citizens were so scattered, that they had no means of acting in concert with each other." It should be a constant injunction to the postmasters, to convey newspapers free of all charge for postage. They are not only the vehicles of knowledge and intelligence, but the sentinels of the liberties of our country.

The conduct of some of those strangers, who have visited our country, since the peace, and who fill the British papers with accounts of our distresses, shows as great a want of good sense, as it does of good nature. They see nothing but the foundations and walls of the temple of liberty; and yet they undertake to judge of the whole fabric.

Our own citizens act a still more absurd part, when they cry out, after the experience of three or four years, that we are not proper materials for republican government. Remember, we assumed these forms of government in a hurry, before we were prepared for them. Let every man exert himself in promoting virtue and knowledge in our country, and we shall soon become good republicans. Look at the steps by which governments have been changed, or rendered stable in Europe. Read the history of Great Britain. Her boasted government has risen out of wars and rebellions that lasted above six hundred years. The United States are traveling peaceably into order and good government. They know no strife—but what arises from the collision of opinions; and, in three years, they have advanced further in the road to stability and happiness, than most of the nations of Europe have done, in as many centuries.

There is but one path that can lead the United States to destruction; and that is their extent of territory. It was probable to effect this, that Great Britain ceded to us so much waste land. But even this path may be avoided. Let but one new state be exposed to sale at a time; and let the land office be shut up, till every part of this new state be settled.

I am extremely sorry to find a passion for retirement so universal among the patriots and heroes of the war. They resemble skillful mariners who, after exerting themselves to preserve a ship from sinking in a storm, in the middle of the ocean, drop asleep as soon as the waves subside, and leave the care of their lives and property, during the remainder of the voyage, to sailors without knowledge or experience. Every man in a republic is public property. His time and talents—his youth—his manhood—his old age—nay more, his life, his all, belong to his country.

Patriots of 1774, 1775, 1776—heroes of 1778, 1779, 1780! come forward! your country demands your services!—Philosophers and friends of mankind, come forward! your country demands your studies and speculations! Lovers of peace and order, who declined taking part in the late war, come forward! your country forgives your timidity and demands your influence and advice! Hear her proclaiming, in sighs and groans, in her governments, in her finances, in her trade, in her manufactures, in her morals, and in her manners, "THE REVOLUTION IS NOT OVER!"

<hr />

APPENDIX C
PROSPECTUS OF A NATIONAL INSTITUTION TO BE ESTABLISHED IN THE UNITED STATES
By Joel Barlow, 1806

[*Reprinted from a defective copy of Barlow's pamphlet in the Congressional Library, supplemented by the reprint in the National Intelligencer of 1806, and a manuscript copy in the possession of Doctor J. C. Welling.*]

The project for erecting a university at the seat of the federal government is brought forward at a happy moment, and on liberal principles. We may therefore reasonably hope for an extensive endowment from the munificence of individuals, as well as from government itself. This expectation will naturally lead us to enlarge our ideas on the subject, and to give a greater scope to its practical operation than has usually been contemplated in institutions of a similar nature.

Two distinct objects, which, in other countries have been kept asunder, may and ought to be united; they are both of great national importance; and by being embraced in the same Institution they will aid each other in their acquisition. These are the advancement of knowledge by association of scientific men, and the dissemination of its rudiments by the instruction of youth. The first has been the business of learned corporations, such as the Royal Society of London and the National Institute of France; the second is pursued by collections of instructors, under the name of universities, colleges, academies, etc.

The leading principle of uniting these two branches of improvement in one Institution, to be extended upon a scale that will render it truly national, requires some development. We find ourselves in possession of a country so vast as to lead the mind to anticipate a scene of social intercourse and interest unexampled in the experience of mankind. This territory presents and will present such a variety of productions, natural and artificial, such a diversity of connections abroad, and of manners, habits, and propensities at home, as will create a strong tendency to diverge and separate the views of those who shall inhabit the different regions within our limits.

It is most essential to the happiness of the people and to the preservation of their republican principles, that this tendency to a separation should be over-balanced by superior motives to a harmony of sentiment; that they may habitually feel that community of interest on which their federal system is founded. This desirable object is to be attained, not only by the operations of the government in its several departments, but by those of literature, sciences, and arts. The liberal sciences are in their nature republican; they delight in reciprocal communication; they cherish fraternal feelings, and lead to a freedom of intercourse, combined with the restraints of society, which contribute together to our improvement.

To explore the natural productions of our country, give an enlightened direction to the labors of industry, explain the advantages of interior tranquility, of moderation and justice in the pursuits of self-interest, and to promote, as far as circumstances will admit, an assimilation of civil regulations, political principles, and modes of education, must engage the solicitude of every patriotic citizen; as he must perceive in them the necessary means of securing good morals and every republican virtue; a wholesome jealousy of right and a clear understanding of duty; without which, no people can be expected to enjoy the one or perform the other for any number of years.

The time is fast approaching when the United States, if no foreign disputes should induce an extraordinary expenditure of money, will be out of debt. From that time forward, the greater part of their public revenue may, and probably will, be applied to public improvements of various kinds; such as facilitating the intercourse through all parts of their dominion by roads, bridges, and canals; such as making more exact surveys, and forming maps and charts of the interior country, and of the coasts, bays, and harbors, perfecting the system of lights, buoys, and other nautical aids; such as encouraging new branches of industry, so far as may be advantageous to the public, either by offering premiums for discoveries, or by purchasing from their proprietors such inventions as shall appear to be of immediate and general utility, and rendering them free to the citizens at large; such as exploring the remaining parts of the wilderness of our continent, both within and without our own jurisdiction, and extending to their savage inhabitants, as far as may be practicable, a taste for civilization, and the means of knowing the comforts that men are capable of yielding to each other in the peaceable pursuits of industry, as they are understood in our stage of society.

To prepare the way for the government to act on these great objects with intelligence, economy, and effect, and to aid its operations when it shall be ready to apply its funds to that purpose, will occupy in part the attention of that branch of the Institution composed of men of scientific research; whose labors, it is expected, will be in great measure gratuitous. It cannot be too early, even at this moment, to direct the researchers of science to occupations of this nature. By these means, at the end of the eleven years, the epoch at which the government may expect to be free of debt, the way can be prepared to begin with system, and proceed with regularity in the various details of public improvement; a business which, if the rulers of all nations did but know it, ought to be considered among the first of their duties, one of the principal objects of their appointment.

The science of political economy is still in its infancy; as indeed is the whole science of government, if we regard it as founded on principles analogous to the nature of man, and designed to promote his happiness. As we believe our government to be founded on these principles, we cannot but perceive an immense field of improvement opening before us; a field in which all the physical as well as the moral sciences should lend their aid and unite their operation, to place human society on such a footing in this great section of the habitable world, as to secure it against further convulsions from violence and war. Mankind have a right to expect this example from us; we alone are in a situation to hold it up before them, to command their esteem, and perhaps their imitation. Should we, by a narrowness of views, neglect the opportunity of realising so many benefits, we ought to reflect that it never can occur to us again; nor can we foresee that it will return to any age or nation. We should grievously disappoint the expectations of all good men in other countries, we should ourselves regret our error while we live; and if posterity did not load us with the reproaches we should merit, it would be because our conduct will have kept them ignorant of the possibility of obtaining the blessings of which it had deprived them.

It would be superfluous, in this Prospectus, to point out the objects merely scientific, that will naturally engage the attention of this branch of the Institution. We are sensible that many of the sciences, physical as well as moral, are very little advanced; some of them, in which we seem to have made considerable progress, are yet so uncertain as to leave it doubtful whether even their first principles do not remain to be discovered; and in all of them, there is a great deficiency as to the mode of familiarizing their results, and applying them to the useful arts of life, the true object of all labor and research.

What a range is open in this country for mineralogy and botany! How many new arts are to arise, and how far the old ones are to be advanced, by the pursuit of these two sciences, it is impossible even to imagine. Chemistry is making a rapid and useful progress, though we still dispute about its elements. Our knowledge of anatomy has laid a necessary and sure foundation for surgery and medicine; surgery indeed is making great proficiency; but, after three thousand years of recorded experience, how little do we know of medicine! Mechanics and hydraulics are progressing fast, and wonderful are the facilities and comforts we draw from them; but while it continues to be necessary to make use of animal force to move heavy bodies in any direction by land or water, we have a right to anticipate new discoveries. Could the genius of a Bacon place itself on the high ground of all the sciences in their present state of advancement, and marshal them before him in so great a country as this, and under a government like ours, he would point out their objects, foretell their successes, and move them on their march, in a manner that should animate their votaries and greatly accelerate their progress.

The mathematics, considered as a science, may probably be susceptible of higher powers than it has yet attained; considered as the handmaid of all the sciences and all the arts, it doubtless remains to be simplified. Some new processes, and perhaps new modes of expressing quantities and numbers, may yet be discovered, to assist the mind in climbing the difficult steps that lead to an elevation so much above our crude conceptions; an elevation that subjects the material universe, with all its abstractions of space and time, to our inspection; and opens, for their combinations, so many useful and satisfying truths.

Researches in literature, to which may be united those in morals, government, and laws, are so vague in their nature, and have been so little methodised, as scarcely to have obtained the name of sciences. No man has denied the importance of these pursuits; though the English nation, from whom we have borrowed so many useful things, has not thought proper to give them that consistency and standing among the objects of laudable ambition, to which they are entitled. Men the most eminent in these studies have not been members of their learned associations. Locke, Berkely, Pope, Hume, Robertson, Gibbon, Adam Smith, and Blackstone, were never admitted into the Royal Society. This is doubtless owing to the nature of their government; though the government itself exerts no influence in these elections. The science of morals connects itself so intimately with the principles of political institutions, that where it is deemed expedient to keep the latter out of sight, it is not strange that the former should meet no encouragement.

This policy is strikingly exemplified in the history of the French Institute. That learned and respectable body was incorporated by the national convention in the year 1795, and took place of all the old academies, which had been previously abolished. It was composed of three classes, according to the objects to be pursued by its members. The first was the class for the physical sciences, the second was the class for the moral and political sciences, the third was for the fine arts. Thus it went on and made great progress in its several branches, till the year 1803, when Bonaparte's government assumed that character which rendered the pursuit of moral and political science inconvenient to him. He then new modeled the Institute, and abolished that class. But lest his real object should be perceived, and he be accused of narrowing the compass of research, he created two new classes in the room of this; one for ancient literature, and one for the French language. On the same occasion an order was issued to all the colleges and great schools in France, suppressing the professorships of moral and political philosophy.

But in our country, and at this early epoch in the course of republican experiment, no subjects of research can be more important than those embraced by these branches of science. Our representative system is new in practice, though some theories of that sort have been framed by speculative writers; and partial trials have been made in the British dominions. But our federal system, combined with democratical representation is a magnificent stranger upon earth; a new world of experiment, bursting with incalculable omens on the view of mankind. It was the result of circumstances which no man could foresee, and no writer pretended to contemplate. It represented itself to us from the necessity of the situation we were in; dreaded at first as an evil by many good men in our country, as well as by our friends in Europe; and it is at this day far from being understood, or properly appreciated, by the generality of those who admire it. Our practice upon it, as far as we have gone, and the vast regions of our continent that present themselves to its embrace, must convince the world that it is the greatest improvement in the mechanism of government that has ever been discovered, the most consoling to the friends of liberty, humanity, and peace.

Men who have grown old in the intrigues of cabinets, and those who, in the frenzy of youthful ambition, present themselves on the theatre of politics, at the head of armies, which they cannot live without, are telling us that "no new principle of government has been discovered for these two thousand years;" [1] and that all proposals to ameliorate the system are vain abstractions, unworthy of sound philosophy. They may tell us too that no new principle in mechanics had been discovered since we came to the knowledge of the lever; no new principle in war, since we first found that a man would cease to fight the moment he was killed. Yet we see in the two latter cases that new combinations of principle have been discovered; they are daily now discovered and carried into practice. In these there are no books written to inform us we can go no further; no imperial decrees to arrest our progress. Why, then, should this be the case in those combinations of the moral sense of man, which compose the science of government?

But whether we consider the principles themselves as new, or the combinations only as new, the fact with respect to our government is this: although the principle has long since been known that the powers necessarily exercised in the people at large, and that these powers cannot conveniently be exercised by the people at large, yet it was not discovered how these powers could be conveniently exercised by a few delegates, in such a manner as to be constantly kept within the reach of the people at large, so as to be controlled by them without a convulsion. But a mode of doing this has been discovered in latter years, and is now for the first time carried into practice in our country; I do not say in the utmost perfection of which the principle is capable; yet in a manner which greatly contributes, with our other advantages, to render us the happiest people on earth. Again, although the principle has long since been known, that good laws faithfully executed within a state, would protect the industry of men, and preserve interior tranquility; yet no method was discovered which would effectually preserve exterior tranquility; yet no method was discovered which would preserve exterior tranquility between state and state. Treaties were made, oaths were exacted, the name of God was invoked, forts, garrisons, and armies were established on their respective frontiers; all with the sincere desire, no doubt, of preserving peace. The whole of these precautions have been constantly found ineffectual. But we at last, and almost by accident, have discovered a mode of preserving peace among states without any of the old precautions; which were always found extremely expensive, destructive to liberty, and incapable of securing the object. We have found that states have some interests that are common and mutual among themselves; that, so far as these interests go, the states should not be independent; that, without losing anything of their dignity, but rather increasing it, they can bind themselves together by a federal government, composed of their own delegates, frequently and freely elected, to whom they can confide these common interests; and that by giving up to these delegates the exercise of certain acts of sovereignty, and retaining the rest to themselves, each state puts it out of its own power to withdraw from the confederation, and out of the power of the general government to deprive them of the rights they have retained.

If these are not new principles of government, they are at least new combinations of principles, which require to be developed, studied, and understood better than they have been, even by ourselves; but especially by the rising generation, and by all foreign observers who shall study our institutions. Foreigners will thus give us credit for what we have done, point out to our attention what we have omitted to do, and perhaps aid us with their lights, in bringing towards perfection a system, which may be destined to ameliorate the condition of the human race.

It is in this view that moral and political research ought to be regarded as one of the most important objects of the National Institution, the highest theme of literary emulation, whether in prose or verse, the constant stimulus to excite the ambition of youth in the course of education.

What are called the fine arts, in distinction from what are called the useful, have been but little cultivated in America. Indeed, few of them have yet arrived, in modern times, to that degree of splendor which they had acquired among the ancients. Here we must examine an opinion, entertained by some persons, that the encouragement of the fine arts savors too much of luxury, and is unfavorable to republican principles. It is true, as is alleged, they have usually flourished most under despotic governments; but so have corn and cattle. Republican principles have never been organized or understood, so as to form a government, in any country but our own. It is therefore from theory, rather than example, that we must reason on this subject. There is no doubt but that the fine arts, both in those who cultivate and those who only admire them, open and expand the mind to great ideas. They inspire liberal feelings, create a harmony of temper, favorable to a sense of justice and a habit of moderation in our social intercourse. By increasing the circle of our pleasures, they moderate the intensity with which pleasures, not dependent on them, would be pursued. In proportion as they multiply our wants, they stimulate our industry, they diversify the objects of our ambition, they furnish new motives for a constant activity of mind and body, highly favorable to the health of both. The encouragement of a taste for elegant luxuries discourages the relish for luxuries that are gross and sensual, debilitating to the body, and demoralising to the mind. These last, it must be acknowledged, are prevailing in our country; they are perhaps the natural growth of domestic affluence and civil liberty. The government, however mild and paternal, cannot check them by any direct application of its powers, without improper encroachments on the liberty and affluence, that give them birth. But a taste for the elegant enjoyments which spring from the culture of the fine arts, excites passions not so irresistible, but that they are easily kept within the limits, which the means of each individual will prescribe. It is the friend of morals and of health; it supposes a certain degree of information; it necessitates liberal instruction; it cannot but be favorable to republican manners, principles, and discipline.

A taste for these arts is peculiarly desirable in those parts of our country, at the southward and westward, where the earth yields her rich productions with little labor, and leaves to the cultivator considerable vacancies of time and superfluities of wealth, which otherwise will, in all probability, be worse employed. The arts of drawing, painting, statuary, engraving, music, poetry, ornamental architecture, and ornamental gardening, would employ a portion of the surplus time and money of our citizens; and at the same time be more likely to dispose their minds to devote another portion to charitable and patriotic purposes, than if the first portion had not been thus employed.

In England there is a Royal Academy for the fine arts, as well as a Royal Society for the sciences; though men of merit in other learned labors are not associated. In France the two classes of eminent men who pursue the sciences and the arts, are united in the National Institute. Besides those, and besides the colleges and universities, there exists in each of those countries a variety of institutions useful in their different objects, and highly conducive to the general mass of public improvement, as well as to private instruction.

The French Government supports,

1. *The School of Mines*, an extensive establishment; where is preserved a collection of specimens from all the mines, wrought and unwrought, that are known to exist in that country; where, with the free use of a laboratory, lectures are given gratis one day in the week for nine months in the year, and where young men receive what is called a mineralogical education. At this place the proprietor of a mine, whether of metals, coals, or other valuable fossils, may have them examined without expense; and here he can apply for an able and scientific artist, recommended by the professors, to be the conductor of his works, as well in the engineering as the metallurgical branch.

2. *The School of Roads and Bridges*; whose title ought to extend likewise to canals, river navigation, and hydraulic architecture; since it embraces all these objects. Here are preserved models and drawings of all the great works, and many of the abortive attempts, in these branches of business. It is a curious and useful collection. This establishment, too, maintains its professors, who give lectures gratis, and produce among their pupils the ablest draftsmen and civil engineers, ready to be employed where the public service or private enterprise may require.

3. *The Conservatory of Arts*; meaning the useful arts and trades. This, in appearance, is a vast Babel of materials; consisting of tools, models, and entire machines, ancient and modern, good and bad. For it is often useful to preserve for inspection a bad machine. The professor explains the reason why it did not answer the purpose; and this either prevents another person from spending his time and money in pursuit of the same impracticable scheme, or it may lead his mind to some ingenious invention to remedy the defect and make it a useful

object. Here is a professor for explaining the use of the machines, and for aiding the minister in discharging the duties of the patent office. Here likewise several trades are carried on, and persons are taught gratis the use of the tools by practice as well as by lectures.

4. *The Museum of Natural History*. This consists of a botanical garden, an extensive menagery, or collection of wild animals, and large cabinets of minerals. To this institution are attached several professorships, and lectures are given on every branch of natural history.

5. *The Museum of Arts*; meaning the fine arts. This is the school for painting, statuary, music, etc. The great splendor of this establishment consists chiefly in its vast gallery of pictures, and its awful synod of statues. These are as far beyond description as they are above comparison. Since, to the collections of the kings of France, the government has added so many of the best productions of Italy, Flanders, and Holland, there is no other assemblage of the works of art where students can be so well accommodated with variety and excellence, to excite their emulation and form their taste.

6. *The National Library*. This collection is likewise unparalleled both for the number and variety of works it contains; having about five hundred thousand volumes, in print and manuscript; besides all of value that is extant in maps, charts, engravings; and a museum of coins, medals, and inscriptions, ancient and modern.

8. *The Mint*; which is a scientific, as well as a laboratorial establishment; where lectures are given in mineralogy, metallurgy, and chemistry.

9. *The Military School*, where field engineering, fortification, gunnery, attack and defence of places, and the branches of mathematics, necessary to these sciences, are taught by experienced masters.

10. *The Prytaneum*; which is an excellent school of general science, more especially military and nautical; but it is exclusively devoted to what are called *enfans de la patrie*, children of the country, or boys adopted by the government, and educated at the public expense. They are generally those whose fathers have died in the public service. But this distinction is often conferred on others, through particular favor. The school is supplied with able instructors; and the

pupils are very numerous. They are taught to consider themselves entirely devoted to the service of their country, as is indicated both by their own appellation and that of their seminary.

11. *The College of France* retains all its ancient advantages, and has been improved by the revolution.

12. *The School of Medicine*, united with anatomy and surgery, is in able hands,
and well conducted.

13. *The Veterinary School*; where practical and scientific lessons are given on
the constitution and diseases of animals.

14. *The Observatory* is an appellation still retained by an eminent school of
astronomy; though its importance has grown far beyond what is indicated by its
name. It publishes the annual work called *la connaissance des tems*; a work not
only of national, but of universal utility for navigators and astronomers.

15. Another institution, whose functions have outgrown its name, is the
Bureau of Longitude. It not only offers premiums for discoveries, tending to the
great object of finding an easy method of ascertaining the longitude at sea, and
judges of their merit; but it is the encourager and depository of all nautical
and geographical discoveries; and, in conjunction with the school of astronomy
and that of natural history, it directs and superintends such voyages of discovery
as the government chooses to undertake.

16. The last public establishment for liberal instruction, that I shall mention
in the capital, though not the only remaining one that might be named, is the
Polytechnic School. This, for the variety of sciences taught, the degree of previous
attainment necessary for admission, the eminent talents of the professors, and
the high state of erudition to which the pupils are carried, is doubtless the first
institution in the world.

The Prytaneum, the Polytechnic School, the Museum of Arts, the Conservatory of Arts, and the Veterinary School, are new institutions, established during the revolution. The others existed before; but most of them have been much
improved. There were likewise erected during the same period, a great number
of provincial colleges. The general provision was to have one in each county, or
department, of which there are upwards of a hundred in France. The provision

likewise extended to what are called primary schools, to be erected and multiplied in every town and village. This is also executed in part, but not completely.

On the whole, the business of education in France is on a much better footing at present than it ever was before the revolution. The clamor that was raised by the emigrants against the convention, reproaching them with having destroyed education, were unfounded; and, we may almost say, the reverse of truth. Their plans on this subject were great, and in general good; much good, indeed, has grown out of them; though they have not been pursued by the government during its subsequent changes, in the manner contemplated by the projectors.

Besides the public foundations, established and partly supported by the government, there is a variety of private associations for collecting and diffusing information; such as agricultural societies, a society for the encouragement of arts and manufactures; and another which, though neither scientific nor literary, is a great encourager of literature. It is a charitable fund for giving relief to indigent authors, and to their widows and orphans.

The Lyceum of Arts, as a private society, merits a distinguished place in this hasty review of the liberal establishments in Paris. This foundation belongs to a number of proprietors, who draw no other advantage from it than the right of attending the lectures, and of using the laboratory, reading rooms, library, and philosophical apparatus. It employs able professors in all the sciences, in technology, in literature, and in several modern languages. It admits annual subscribers, who enjoy these advantages during the year; and it is particularly useful to strangers and to young men from the provinces, who might otherwise employ their leisure hours in less profitable amusements.

If, in speaking of the state of public instruction in England, we are less particular than in those of her neighbors, it will not be for want of respect for her institutions; but because most of them are better known in this country, and some of them similar to those we have described. Her universities and colleges, her numerous agricultural societies, her society of arts and manufactures, her royal society, royal academy, royal observatory, British museum, marine and military academies, her society for exploring the interior of Africa, her missionary society, and her board of longitude, are probably familiar to most of the readers of this Prospectus. We shall particularise only two or three others; which, being of recent date, are probably less known.

The Literary Fund, for the relief of indigent authors and their families, is an institution of extensive and increasing beneficence. It is not merely a charitable, but a patriotic endowment; and its influence must extend to other nations, and to posterity. For an author of merit belongs to the world at large; his genius is not the property of one age or nation, but the general heritage of all. When a fund like this is administered by men of discernment and fidelity, worthy of their trust, as the one in question certainly is, lending its aid to all proper objects, without regard to party or system, whether in politics, science, or religion, it gives independence to literary pursuits. Men who are fostered by it, or feel a confidence that they may, in case of need, partake of its munificence, become bold in the development of useful truths; they are not discouraged by the dread of opposing the opinions of vulgar minds, whether among members of the government, or powerful individuals.

This generous and energetic establishment owes it foundation to David Williams; whose luminous writings, as well as other labors, in favor of liberty and morals, are well known in this country. It was a new attempt to utilize the gifts of fortune, and the efforts of timid merit. It was not till after many years of exertion by its patriotic founder, that the institution assumed a vigorous existence, became rich by the donations of the opulent, and popular from the patronage of the first names in the kingdom. It was from this fund that the one of a similar nature in Paris was copied; but the latter is hitherto far inferior to the former, both in its endowments and its activity.

On the other hand, the *Royal Institution* and the *London Institution* have been copied from the Lyceum in Paris. But in these instances the copies have already equaled, if not surpassed, the original.

We have traced this rapid sketch of what is doing for the advancement of liberal knowledge and public improvements in other countries, for the sake of grouping the whole in one general view; that we may compare their establishments with our situation, our wants, our means, and our prospects; reject what is unsuitable to us, adopt such as would be useful, and organize them as shall be advantageous in our National Institution.

It is proposed, as already observed, that this Institution should combine the two great objects, *research* and *instruction*. It is expected from every member that he will employ his talents gratuitously in contributing to the *first* of these objects. The *second* will be the special occupation of a branch of the Institution, to be styled the Professorate. And, as it is expected from the members of this branch, that they devote their time as well as talents to the labor of instruction, they will receive a suitable compensation, to be fixed by the board of trustees.

The members of the National Institution shall be elected from citizens of the United States, eminent in any of the liberal sciences, whether physical, moral, political, or economical; in literature, arts, agriculture; in mechanical, nautical, or geographical discoveries. The number of members shall at no time exceed the decuple of the number of states, composing the confederation of the United States. But in addition to these, it may elect honorary members abroad, not exceeding in number one-half of that of its members. And it may likewise elect corresponding members within the United States, or elsewhere, not exceeding the last-mentioned proportion.

The members of the Institution may divide themselves into several sections, for their more convenient deliberations on the objects of their several pursuits, not exceeding five sections. Each section shall keep a register of its proceedings. It shall be the duty of each section to nominate candidates for members of the Institution, suitable for each section. Which nomination, if there be vacancies, shall entitle such candidates to be balloted for at the general meetings.

There shall be a Chancellor of the National Institution; whose duty it shall be to superintend its general concerns. He shall, in the first instance, be appointed by the President of the United States; and hold his office during the pleasure of the Institution. He shall preside in its general meetings; direct the order of its deliberations, and sign the diplomas of its members. He shall be president of the board of trustees; and, in consequence of their appropriations, order the payment of monies, and otherwise carry into execution their ordinances and resolutions. He shall be director of the Professorate; order the courses of lectures and other modes of instruction and objects of study; confer degrees in the central university; appoint examiners, either at the district colleges or at the central university, for the admission of students into the latter; fill vacancies in the Professorate, until the next meeting of the board of trustees; and he shall have power to suspend from office a professor, until the time of such meeting. He shall instruct and direct in their mission, such traveling professors as the board of trustees shall employ, for the objects of science, in our own country or abroad.

The board of trustees shall consist of fifteen members; they shall be first appointed by the President of the United States, and hold their office during the pleasure of the Institution. They shall give bonds with surety for the faithful execution of their trust. They and the chancellor are, of course, members of the Institution. As soon as convenient after their appointment, they are to assemble at the seat of government, elect by ballot fifteen additional members of the Institution, appoint three professors, and transact such other business as they may think proper. But no more than the second fifteen members of the Institution shall be elected, until the last Wednesday in November next. On which day a general meeting of the Institution shall be held at the seat of government; and the members then present may proceed to elect fifteen additional members. Two months after which, another election of fifteen members may take place; but no more until the November then next. Thus they may proceed to hold two elections in each year, of fifteen members each, if they think proper, till the whole number allowed by law shall be elected. The Institution will fill its own vacancies, and those in the board of trustees, appoint its treasurer and secretaries and, on all occasions after the first, elect the chancellor.

The chancellor and board of trustees shall have the sole management of the funds of the Institution, whether in lands or movables; they shall organize the Professorate, appoint the professors and other masters and teachers; assign them their compensations, and remove them at pleasure. They shall establish a central university at or near the seat of government, and such other universities, colleges, and schools of education, as the funds of the Institution will enable them to do, whether in the city of Washington, or in other parts of the United

States; and make the necessary regulations for the government and discipline of the same. They may likewise establish printing presses for the use of the Institution, laboratories, libraries, and apparatus for the sciences and the arts; and gardens for botany and agricultural experiments.

Thus organized, and with proper endowments, the National Institution will be able to expand itself to a large breadth of public utility. It will, by its correspondence, its various establishments, its premiums, its gratuities, and other encouragements, excite a scrupulous attention to the duties of education in every part of the United States. By printing school books in the vast quantities that are wanted, and selling them at prime cost, it will furnish them at one-third of the price usually demanded; and by an able selection or composition of such as are best adapted to the purpose, it will give a uniformity to the moral sentiment, a republican energy to the character, a liberal cast to the mind and manners of the rising and following generations. None will deny that these things are peculiarly essential to the people of this country; for the preservation of their republican principles, and especially of their federal system.

Add to this the advantages that the government will draw, in its projected plans of public improvement, from this facility of concentrating the rays of science upon the most useful objects; from directing the researches of so many of the ablest men in the country to the best modes of increasing its productions and its happiness; from having a greater choice of young and well-taught engineers, civil and military; as well as mechanicians, architects, geologists; and men versed in the mathematical sciences and political economy.

Attached to the university in Washington, and under the direction of the Institution, might be the best position for the military academy, now at West Point; as likewise for the naval academy and for the mint of the United States. The patent office is now an embarrassing appendage to the department of state. It might occupy very usefully one of the professors of this university. The machines and models belonging to it would be useful ornaments in a lecture room, where mechanics, hydraulics, and other branches of natural philosophy are taught. Such professor might be the proper person to examine the applications for patents, and report upon their merits; the chancellor might grant the patents. It might likewise be advantageous, that the trustees, when the state of their funds will permit, should purchase from their proprietors such inventions as, in their opinion, might be of immediate and general use; and perhaps the chancellor might be authorized to refuse patents for impracticable things, and expose to public view such imposters as sometimes apply for them, with the intention of imposing upon the credulous, by selling their fallacious privileges either in whole or in part.

The geographical and mineralogical archives of the nation might be better placed in this university than elsewhere. Being confided to professors, they might draw advantages from them in the course of their instructions. Thus the Institution might become a general depositary of the results of scientific research; of experiments in art, manufactures, and husbandry; and of discoveries by voyages and travels. In short, no rudiment of knowledge should be below its attention, no height of improvement above its ambition, no corner of our empire beyond its vigilant activity for collecting and diffusing information.

It is hoped that the legislature, as well as our opulent citizens, will assist in making a liberal endowment for so great an object, and as soon as circumstances will admit; as too much time has already been lost since the government has taken its definitive stand, in so advantageous a position for the development of this part of our national resources.

APPENDIX

Such is the outline of a system of Public Instruction, that would seem to promise the greatest benefits. And, although under present circumstances, it is doubtless too extensive to be carried into immediate practice in all its parts; yet there are strong reasons to wish that its general basis may be preserved entire, in the law for incorporating the Institution; and that such law may be enacted during the present session of Congress. Believing that no possible disadvantage could arise from adopting both of these propositions, we will endeavor to elucidate the advantages by a few additional observations.

1. As we must solicit donations from individual citizens, and depend princi-

pally on them for its endowment, we ought to have a basis on which they can repose their confidence. This can only be done by a board of trustees, standing on the ground of a corporation; whose object is clearly defined; and which is composed of men of known character and responsibility, anxious themselves to promote the object, and pledged in honor and reputation for its ultimate success.

2. The present appears to be a more favorable moment for an establishment of this kind, and especially for obtaining donations, than can be expected to arrive hereafter. A general opinion now prevails, that education has been too much and too long neglected in most parts of our country; and this opinion is happily accompanied by a liberal spirit on the subject; a spirit worthy of the age and country in which we live, and of the government that conducts our affairs. It is a patriotic spirit, that only requires to be directed; but if not directed, may soon be lost. The opinions and dispositions of men are changeable. The race of patriarchs who framed our political systems, and are peculiarly solicitous to ensure their permanent support, are passing off the stage of public life. Children are growing up, to take the legacy we are bequeathing them, insensible of its value, and ignorant of the means by which it can be preserved. It will seem as if we had labored in vain, if we leave our work but half accomplished. And surely the task of preserving liberty, if not as bold, is at least as difficult, as that of acquiring it.

To acquire liberty, comparatively speaking, is the work of few; to preserve it is the sober and watchful business of all. In the first operation, a group of well-informed, enthusiastic, and patriotic leaders, step forward to the field of danger, impress their own energy on the multitude of followers; who cannot go wrong, because the object is palpable, and clearly understood; but in the second, the impetuosity of enthusiasm is no longer the weapon to be used; the mass of the people are masters; they must be instructed in their work; and they may justly say, that when their leaders taught them how to gain their liberty, they contracted the obligation to teach them how to use it.

3. The Institution, though established on the broad foundation we here propose, will begin upon a small scale; no larger than its means will render convenient. And the magnitude of the perspective will not discourage its infant ex-

ertions, but rather increase them. Its expenditures will not be greater at the beginning than they would be if it were always to be confined to the narrow compass in which it will move at first. It will immediately open a few schools at Washington, where they are much wanted. It may soon begin to receive donations for this and other objects; and by its correspondence, it will be learning the wants of the different districts of the United States, and directing its enquiries how to supply them.

4. It is believed that several men of science, without any compensation, but the pleasure of being useful, may be engaged to give courses of lectures during the next winter, on some of the higher branches of knowledge; such as chemistry, mathematics, natural and moral philosophy, political economy, medicine, and jurisprudence: that it may no longer be said of the capital of the United States, that it offers no attractions as a winter residence to strangers or citizens; no amusements but such as are monotonous, and unimproving; nothing to variegate the scene and enliven the labors of those whom the confidence of their country has called to this place, to manage her great concerns. A few courses of lectures on these subjects, announced in the public papers, to be delivered next winter, would draw to this place many young men from the different states; who, being at a loss for the means of finishing their education, are often driven to Europe for that purpose. This would be a beginning for the university, and lead to its interior organisation. It would help to bring the Institution into notice, be the means of augmenting its endowments, and enable the trustees to devise measures for some of their buildings.

5. It ought not to be forgotten that a central Institution of this kind in the United States would not only remove the disadvantages that our young men now experience, in being obliged to obtain a European education; but it would federalize, as well as republicanise their education at home. Coming together from all parts of the union, at an age, when impressions on the mind are not easily effaced, the bent of intellect will attain a similarity in all, diversified only by what nature had done before; their moral characters would be cast in a kindred mould; they would form friendships, which their subsequent pursuits in life would never destroy. This would greatly tend to strengthen the political union of the states, a union which, though founded on permanent interest, can

only be supported by a permanent sense of that interest. In addition to the other advantages of study, we ought to notice the great political school that will be open to the student, during the sessions of Congress; the school of jurisprudence in the federal courts; the constant examples of enlarged ideas, and paternal solicitude for the national welfare, which he will see in the several departments of the executive government.

When the men, who shall have finished their education in this central seat, shall return to it in maturer life, clothed with the confidence of their fellow citizens, to assist in the councils of the nation, the scene will enliven the liberal impressions of youth, combined with the cautious that experience will have taught. They will bring from home the feelings and interests of their own districts; and they will mingle them here with those of the nation. From such men the Institution may perceive the good it may have done; and from them it will learn what new openings may be found in the different states, for the extension of its benefits.

WASHINGTON, *24th January, 1806.*

APPENDIX D
THE MORRILL ACT[1]

AN ACT donating Public Lands to the several States and Territories which may provide Colleges for the Benefit of Agriculture and Mechanic Arts.

Be it enacted by the Senate and House of Representatives of the United States of America in Congress assembled, That there be granted to the several States, for the purposes hereinafter mentioned, an amount of public land, to be apportioned to each State a quantity equal to thirty thousand acres for each senator and representative in Congress to which the States are respectively entitled by the apportionment under the census of eighteen hundred and sixty: *Provided,* That no mineral lands shall be selected or purchased under the provisions of this act.

SEC. 2. *And be it further enacted,* That the land aforesaid, after being surveyed, shall be apportioned to the several States in sections or subdivisions of sections, not less than one-quarter of a section; and whenever there are public lands in a State subject to sale at private entry at one dollar and twenty-five cents per acre, the quantity to which said State shall be entitled shall be selected from such lands within the limits of such State, and the Secretary of the Interior is hereby directed to issue to each of the States in which there is not the quantity of public lands subject to sale at private entry at one dollar and twenty-five cents per acre, to which said State may be entitled under the provisions of this act, land scrip to the amount in acres for the deficiency of its distributive share: said scrip to be sold by said States and the proceeds thereof applied to the uses and purposes prescribed in this act, and for no other use or purpose whatsoever; *Provided*, That in no case shall any State to which land scrip may thus be issued be allowed to locate the same within the limits of any other State, or of any Territory of the United States, but their assignees may thus locate said land scrip upon any of the unappropriated lands of the United States subject to sale at private entry at one dollar and twenty-five cents, or less, per acre: *And provided, further*, That not more than one million acres shall be located by such assignees in any one of the States: *And provided, further*, That no such location shall be made before one year from the passage of this act.

SEC. 3. *And be it further enacted*, That all the expenses of management, superintendence, and taxes from date of selection of said lands, previous to their sales, and all expenses incurred in the management and disbursement of the moneys which may be received therefrom, shall be paid by the States to which they may belong, out of the treasury of said States, so that the entire proceeds of the sale of said lands shall be applied without any diminution whatever to the purposes hereinafter mentioned.

SEC. 4. *And be it further enacted*, That all moneys derived from the sale of the lands aforesaid by the States to which the lands are apportioned, and from the sales of land scrip hereinbefore provided for, shall be invested in stocks of the United States, or of the States, or some other safe stocks yielding not less than five per centum upon the par value of said stocks; and that the moneys so invested shall constitute a perpetual fund, the capital of which shall remain forever undiminished, (except so far as may be provided in section fifth of this act,) and the interest of which shall be inviolably appropriated, by each State which may take and claim the benefit of this act, to the endowment, support, and maintenance of at least one college where the leading object shall be, without excluding other scientific and classical studies, and including military tactics, to teach such branches of learning as are related to agriculture and the mechanic arts, in such manner as the legislatures of the States may respectively prescribe, in order to promote the liberal and practical education of the industrial classes in the several pursuits and professions in life.

SEC. 5. *And be it further enacted,* That the grant of land and land scrip hereby authorized shall be made on the following conditions, to which, as well as to the provisions hereinbefore contained, the previous assent of the several States shall be signified by legislative acts:

First. If any portion of the fund invested, as provided by the foregoing section, or any portion of the interest thereon, shall, by any action or contingency be diminished or lost, it shall be replaced by the State to which it belongs, so that the capital of the fund shall remain forever undiminished; and the annual interest shall be regularly applied without diminution to the purposes mentioned in the fourth section of this act, except that a sum, not exceeding ten per centum upon the amount received by any State under the provisions of this act, may be expended for the purchase of lands for sites or experimental farms, whenever authorized by the respective legislatures of said States.

Second. No portion of said fund, nor the interest thereon, shall be applied, directly or indirectly, under any pretence whatever, to the purchase, erection, preservation, or repair of any building or buildings.

Third. Any State which may take and claim the benefit of the provisions of this act shall provide, within five years, at least not less than one college, as described in the fourth section of this act, or the grant to such State shall cease; and said State shall be bound to pay the United States the amount received of any lands previously sold, and that the title to purchasers under the State shall be valid.

Fourth. An annual report shall be made regarding the progress of each college, recording any improvements and experiments made, with their costs and results, and such other matters, including State industrial and economical statistics, as may be supposed useful; one copy of which shall be transmitted by mail free, by each, to all the other colleges which may be endowed under the provisions of this act, and also one copy to the Secretary of the Interior.

Fifth. When lands shall be selected from those which have been raised to double the minimum price, in consequence of railroad grants, they shall be computed to the States at the maximum price, and the number of acres proportionally diminished.

Sixth. No State while in a condition of rebellion or insurrection against the government of the United States shall be entitled to the benefit of this act.

Seventh. No State shall be entitled to the benefits of this act unless it shall express its acceptance thereof by its legislature within two years from the date of its approval by the President.

SEC. 6. *And be it further enacted*, That land scrip issued under the provisions of this act shall not be subject to location until after the first day of January, one thousand eight hundred and sixty-three.

SEC. 7. *And be it further enacted*, That the land officers shall receive the same fees for locating land scrip issued under the provisions of this act as is now allowed for the location of military bounty land warrants under existing laws: *Provided*, Their maximum compensation shall not be thereby increased.

SEC. 8. *And be it further enacted*, That the Governors of the several States to which scrip shall be issued under this act shall be required to report annually to Congress all sales made of such scrip until the whole shall be disposed of, the amount received for the same, and what appropriation has been made of the proceeds.

Approved, July 2, 1862.

THE HATCH ACT[2]

[*Forty-ninth Congress, second session, chapter 314. Statutes of the United States, Vol. XXIV, page 440.*]

AN ACT to establish agricultural experiment stations in connection with the colleges established in the several states under the provisions of an act approved July second, eighteen hundred and sixty-two, and of the acts supplementary thereto.

Be it enacted by the Senate and House of Representatives of the United States of America in Congress assembled, That in order to aid in acquiring and diffusing among the people of the United States useful and practical information on subjects connected with agriculture, and to promote scientific investigation and experiment respecting the principles and applications of agricultural science, there shall be established, under direction of the college or colleges or agricultural department of colleges in each State or Territory established, or which may hereafter be established, in accordance with the provisions of an act approved July second, eighteen hundred and sixty-two, entitled "An act donating public lands to the several States and Territories which may provide colleges for the benefit of agriculture and the mechanic arts," or any of the supplements to said act, a department to be known and designated as an "agricultural experiment station:" *Provided,* That in any State or Territory in which two such colleges have been or may be so established the appropriation hereinafter made to such State or Territory shall be equally divided between such colleges, unless the legislature of such State or Territory shall otherwise direct.

SEC. 2. That it shall be the object and duty of said experiment stations to conduct original researchers or verify experiments on the physiology of plants and animals; the diseases to which they are severally subject, with the remedies for the same; the chemical composition of useful plants at their different stages of growth; the comparative advantage of rotative cropping as pursued under a varying series of crops; the capacity of new plants or trees for acclimation; the analysis of soils and water; the chemical composition of manures, natural or artificial, with experiments designed to test their comparative effects on crops of different kinds; the adaptation and value of grasses and forage plants; the composition and digestibility of the different kinds of food for domestic animals; the scientific and economic questions involved in the production of butter and cheese; and such other researches or experiments bearing directly on the agricultural industry of the United States as may in each case be deemed advisable, having due regard to the varying conditions and needs of the respective States or Territories.

SEC. 3. That in order to secure, as far as practicable, uniformity of methods and results in the work of said stations, it shall be the duty of the United States

Commissioner of Agriculture to furnish forms, as far as practicable, for the tabulation of results of investigation or experiments; to indicate, from time to time, such lines of inquiry as to him shall seem most important; and, in general, to furnish such advice and assistance as will best promote the purposes of this act. It shall be the duty of each of said stations, annually, on or before the first day of February, to make to the governor of the State or Territory in which it is located a full and detailed report of its operations, including a statement of receipts and expenditures, a copy of which report shall be sent to each of said stations, to the said Commissioner of Agriculture, and to the Secretary of the Treasury of the United States.

SEC. 4. That bulletins or reports of progress shall be published at said stations at least once in three months, one copy of which shall be sent to each newspaper in the States or Territories in which they are respectively located, and to such individuals actually engaged in farming as may request the same, and as far as the means of the station will permit. Such bulletins or reports and the annual reports of said stations shall be transmitted in the mails of the United States free of charge for postage, under such regulations as the Postmaster-General may from time to time prescribe.

SEC. 5. That for the purpose of paying the necessary expenses of conducting investigations and experiments and printing and distributing the results as hereinbefore prescribed, the sum of fifteen thousand dollars per annum is hereby appropriated to each State, to be specially provided for by Congress in the appropriations from year to year, and to each Territory entitled under the provisions of section eight of this act, out of any money in the Treasury proceeding from the sales of public lands, to be paid in equal quarterly payments, on the first day of January, April, July, and October in each year, to the treasurer or other officer duly appointed by the governing boards of said colleges to receive the same, the first payment to be made on the first day of October, eighteen hundred and eighty-seven: *Provided, however*, That out of the first annual appropriation so received by any station an amount not exceeding one-fifth may be expended in the erection, enlargement, or repair of a building or buildings necessary for carrying on the work of such station; and thereafter an amount not exceeding five per centum of such annual appropriation may be so expended.

SEC. 6. That whenever it shall appear to the Secretary of the Treasury from the annual statement of receipts and expenditures of any of said stations that a portion of the preceding annual appropriation remains unexpended, such amount shall be deducted from the next succeeding annual appropriation to such station, in order that the amount of money appropriated to any station shall not exceed the amount actually and necessarily required for its maintenance and support.

SEC. 7. That nothing in this act shall be construed to impair or modify the legal relation existing between any of the said colleges and the government of the States or Territories in which they are respectively located.

SEC. 8. That in States having colleges entitled under this section to the benefits of this act and having also agricultural experiment stations established by law separate from said colleges, such States shall be authorized to apply such benefits to experiments at stations so established by such States; and in case any State shall have established under the provisions of said act of July second aforesaid, an agricultural department or experimental station, in connection with any university, college, or institution not distinctively an agricultural college or school, and such State shall have established or shall hereafter establish a separate agricultural college or school, which shall have connected therewith an experimental farm or station, the legislature of such State may apply in whole or in part the appropriation by this act made, to such separate agricultural college or school, and no legislature shall by contract express or implied disable itself from so doing.

SEC. 9. That the grants of moneys[3] authorized by this act are made subject to the legislative assent of the several States and Territories to the purposes of said grants: *Provided*, That payment of such instalments of the appropriation herein made as shall become due to any State before the adjournment of the regular session of its legislature meeting next after the passage of this act shall be made upon the assent of the governor thereof duly certified to the Secretary of the Treasury.

SEC. 10. Nothing in this act shall be held or construed as binding the United States to continue any payment from the Treasury to any or all the States or institutions mentioned in this act, but Congress may at any time amend, suspend, or repeal any or all the provisions of this act.

Approved, March 2, 1887.

APPENDIX E
A LIST OF STATE UNIVERSITIES AND FEDERAL LAND-GRANT COLLEGES, WITH THE DATES OF THEIR ORGANIZATION

NOTE.—Most of the State universities owe their origin wholly or in part to Federal land grants in connection with the Morrill act, or by special acts passed by Congress. The thirteen original States and six others have received no land grants, except for agricultural and mechanical colleges. All the Territories have had land grants for educational purposes except the District of Columbia and Alaska. Of the thirteen original States only four—Virginia, Georgia, and North and South Carolina—have founded and maintained State universities; six—Massachusetts, Connecticut, Pennsylvania, New Jersey, Rhode Island, and New Hampshire—founded in colonial days institutions which have become practically State universities; New York, though fairly liberal to its colleges, has never concentrated its patronage; Maryland and Delaware have practically ignored the university question. In the other States without grants—Vermont, Maine, Kentucky, Tennessee, Texas, and West Virginia—the efforts to found State institutions have been attended with much difficulty, and it is evident to one who studies the subject that their educational systems are probably much less prosperous than they would have been had they received assistance from the General Government similar to that given their sister States.[1]

In the following list institutions wholly or in part supported by the State are designated by the symbol †. Institutions organized or extended in scope in connection with the Morrill act of 1862 are designated by the symbol *. Institutions maintained in connection with the Hatch act are designated by the symbol △. Institutions whose names are indented are subordinated to those which precede them.

The total amount of land given by the General Government for State educational work has been 1,995,920 acres. The total amount appropriated by the States for higher education is shown by Blackmar to have been $27,475,646.

I am indebted to Professor F. W. Blackmar, Professor W. O. Atwater, and Mr. A. C. True for the facts embodied in the following tables:

ALABAMA
(Territory, 1817; State, 1819; land grant, 1818–19)
†UNIVERSITY OF ALABAMA, Tuscaloosa, 1819–1821
*ALABAMA AGRICULTURAL AND MECHANICAL COLLEGE,
    Auburn, 1872
△AGRICULTURAL EXPERIMENT STATION, Auburn, 1883
△†CANEBRAKE AGRICULTURAL EXPERIMENT STATION,
    Uniontown, 1885
    Alabama Historical Society, Tuscaloosa, 1851
    No scientific society in the State

ALASKA
(Territory, 1872)
No colleges
Alaska Historical Society, Sitka, 1890
Society of Alaskan Natural History and Ethnology, Sitka, 1887

ARIZONA
(Territory, 1863; land grant, 1881)
UNIVERSITY OF ARIZONA, Tucson, 1889
    COLLEGE OF AGRICULTURE, UNIVERSITY OF ARIZONA, Tucson,
        1889
        No historical or scientific society

ARKANSAS

(Territory, 1819; State 1836; land grant, 1836)

*†ARKANSAS INDUSTRIAL UNIVERSITY, Fayetteville, 1869–1872

△ARKANSAS AGRICULTURAL EXPERIMENT STATION, Fayetteville, 1888

(Substations at Pine Bluff, Newport, and Texarkana)

Arkansas Historical Society, Little Rock

No scientific society

CALIFORNIA

(Territory, 1846; State, 1850; land grant, 1853)

†*UNIVERSITY OF CALIFORNIA, Berkeley, 1868–69

COLLEGE OF AGRICULTURE, MECHANICS, MINING, ENGINEERING, AND CHEMISTRY, UNIVERSITY OF CALIFORNIA, Berkeley, 1866–1868

△AGRICULTURAL EXPERIMENT STATION, UNIVERSITY OF CALIFORNIA, Berkeley, 1876 and 1888

(Outlying stations at Paso Robles, Tulare, Jackson, Cupertino, Fresno, Mission San Jose)

California Historical Society, San Francisco

California Academy of Sciences, San Francisco, 1854

COLORADO

(Territory, 1861; State, 1876; land grant, 1875)

†UNIVERSITY OF COLORADO, Boulder, 1875–1877

*STATE AGRICULTURAL COLLEGE, Fort Collins, 1879

△AGRICULTURAL EXPERIMENT STATION OF COLORADO, Fort Collins, 1888

(Substations at Del Norte and Rocky Ford)

†STATE SCHOOL OF MINES, Golden, 1874

Colorado State Historical Society, Denver

Colorado Scientific Society, Denver

CONNECTICUT
(Settled, 1634; State, 1788)
YALE UNIVERSITY, New Haven, 1700
*SHEFFIELD SCIENTIFIC SCHOOL OF YALE UNIVERSITY, 1847 and
1864
△†CONNECTICUT AGRICULTURAL EXPERIMENT STATION,
New Haven, 1875 and 1877
STORRS AGRICULTURAL SCHOOL, Mansfield, 1881
STORRS SCHOOL AGRICULTURAL EXPERIMENT STATION, 1888
Connecticut Academy of Sciences, New Haven, 1799
Connecticut Historical Society, Hartford, 1825

DAKOTA, NORTH
(Territory of Dakota, 1861; State, 1889; land grant, 1881)
(?) UNIVERSITY OF NORTH DAKOTA, Grand Forks, 1883–4
NORTH DAKOTA AGRICULTURAL COLLEGE, Fargo, 1890
No State historical or scientific society

DAKOTA, SOUTH
(State, 1889; land grant, 1881)
(?) UNIVERSITY OF SOUTH DAKOTA, Vermilion, 1883
†SOUTH DAKOTA AGRICULTURAL COLLEGE, Brookings, 1889
SOUTH DAKOTA AGRICULTURAL EXPERIMENT STATION,
Brookings, 1888
SOUTH DAKOTA SCHOOL OF MINES, Rapid City, 1886
No State historical or scientific society

DELAWARE
(Settled, 1638; State, 1787)
†DELAWARE COLLEGE, Newark, 1834, 1851, and 1871
△DELAWARE COLLEGE AGRICULTURAL EXPERIMENT STATION,
Newark, 1888
Historical Society of Delaware, Wilmington, 1884
No scientific society

FLORIDA
(Territory, 1821; State, 1845; land grant, 1845)
*FLORIDA STATE AGRICULTURAL AND MECHANICAL COLLEGE,
Lake City, 1884
△AGRICULTURAL EXPERIMENT STATION OF FLORIDA, Lake City,
1888
Historical Society of Florida, St. Augustine

GEORGIA
(Settled, 1732; State, 1788)
†*UNIVERSITY OF GEORGIA, Athens [1784], 1801
GEORGIA STATE COLLEGE OF AGRICULTURE AND MECHANIC
ARTS, OF THE UNIVERSITY OF GEORGIA, Athens, 1872
△GEORGIA AGRICULTURAL EXPERIMENT STATION, Athens, 1888
SOUTHWEST GEORGIA AGRICULTURAL COLLEGE, UNIVERSITY
OF GEORGIA, Cuthbert, 1879
NORTH GEORGIA AGRICULTURAL COLLEGE, UNIVERSITY OF
GEORGIA, Dahlonega, 1873
WEST GEORGIA AGRICULTURAL AND MECHANICAL COLLEGE,
Hamilton, 1882
MIDDLE GEORGIA MILITARY AND AGRICULTURAL COLLEGE,
UNIVERSITY OF GEORGIA, Milledgeville, 1880
SOUTH GEORGIA COLLEGE OF AGRICULTURE AND THE ME-
CHANIC ARTS, UNIVERSITY OF GEORGIA, Thomasville, 1879
†ATLANTA UNIVERSITY (colored), Atlanta, 1859
Georgia Historical Society, Savannah, 1839
No scientific society

ILLINOIS
(Territory, 1809; State, 1818; land grants, 1804 and 1818)
†UNIVERSITY OF ILLINOIS, Urbana, 1868. (Formerly Illinois Industrial
University)

*COLLEGE OF AGRICULTURE OF THE UNIVERSITY OF ILLINOIS,
Urbana, 1867
△AGRICULTURAL EXPERIMENT STATION OF THE UNIVERSITY
OF ILLINOIS, Champaign, 1888
Illinois State Historical Society, Champaign
No State scientific study

INDIANA
(Territory, 1800; State, 1816; land grants, 1804 and 1816)
†INDIANA UNIVERSITY, Bloomington, 1820–26. (Successor to Vincennes
University, 1806)
PURDUE UNIVERSITY, Lafayette, 1874
*SCHOOL OF AGRICULTURE, HORTICULTURE, AND VETERINARY
SCIENCE OF PURDUE UNIVERSITY, Lafayette, 1873
△AGRICULTURAL STATION OF INDIANA, Lafayette, 1887
Indiana Historical Society, Indianapolis, 1832
Indiana Academy of Sciences (unlocalized), 1885

IOWA
(Territory, 1838; States, 1846; land grant, 1845)
†STATE UNIVERSITY OF IOWA, Iowa City, 1847–60
*IOWA STATE COLLEGE OF AGRICULTURE AND MECHANIC ARTS,
Ames, 1858; opened for students October 21, 1868
△IOWA AGRICULTURAL EXPERIMENT STATION, Ames, 1888
Iowa State Historical Society, Iowa City
Davenport Academy of Sciences, Davenport, 1867
Iowa Academy of Sciences, Iowa City, 1875

KANSAS
(Territory, 1857; State, 1861, land grant, 1861)
†UNIVERSITY OF KANSAS, Lawrence, 1861–1866
KANSAS STATE AGRICULTURAL COLLEGE, Manhattan, 1863
△KANSAS AGRICULTURAL EXPERIMENT STATION, Manhattan, 1888

Kansas State Historical Society, Topeka
Kansas Academy of Science, Topeka, 1868

KENTUCKY
(State, 1792)
*AGRICULTURAL AND MECHANICAL COLLEGE OF KENTUCKY,
        Lexington, 1865; reorganized, 1880. (Successor to Transylvania
        University, organized 1798)
△KENTUCKY AGRICULTURAL EXPERIMENT STATION, Lexington,
        1885
        Kentucky Historical Society, Frankfort
        No State scientific society

LOUISIANA
(Territory, 1803; State, 1812; land grants, 1806, 1811, 1827)
TULANE UNIVERSITY OF LOUISIANA, New Orleans, 1847
†SOUTHERN UNIVERSITY (colored), New Orleans, 1880
†*LOUISIANA STATE UNIVERSITY AND AGRICULTURAL AND
        MECHANICAL COLLEGE, Baton Rouge, 1873; reorganized, 1877
†△SUGAR EXPERIMENT STATION NO. 1, Kenner, 1885
△†SUGAR EXPERIMENT STATION NO. 2, Baton Rouge, 1886
△†NORTH LOUISIANA EXPERIMENT STATION, Calhoun, 1888
        Louisiana Historical Society, Baton Rouge
        No State scientific society

MAINE
(Settled, 1622; State, 1820)
*MAINE STATE COLLEGE OF AGRICULTURE AND THE MECHANIC
        ARTS, Orono, 1865.[2]
△MAINE STATE COLLEGE AGRICULTURAL EXPERIMENT STATION,
        Orono, 1885 and 1887
        Maine Historical Society, Portland, 1822
        No State scientific society

MARYLAND
(Settled, 1631; State, 1788)

[UNIVERSITY OF MARYLAND, organized 1784; abandoned, 1805]

*MARYLAND AGRICULTURAL COLLEGE,
Agricultural College [1856], 1859

△MARYLAND AGRICULTURAL EXPERIMENT STATION,
Agricultural College, 1888

Maryland Academy of Sciences, 1822

Maryland Historical Society, Baltimore

MASSACHUSETTS
(Settled, 1620; State, 1788)

HARVARD UNIVERSITY, Cambridge, 1636[3]

*MASSACHUSETTS INSTITUTE OF TECHNOLOGY, Boston, 1863–1865

*MASSACHUSETTS AGRICULTURAL COLLEGE, Amherst, 1856, 1863,
and 1867

†MASSACHUSETTS STATE AGRICULTURAL EXPERIMENT STATION,
Amherst, 1882 and 1888

△HATCH EXPERIMENT STATION OF MASSACHUSETTS
AGRICULTURAL COLLEGE, Amherst, 1888

American Academy of Arts and Sciences, 1780

Massachusetts Historical Society, Boston

MICHIGAN
(Territory, 1805; State, 1836; land grant, 1836)

†UNIVERSITY OF MICHIGAN, Ann Arbor [1817], 1836, 1840

*MICHIGAN AGRICULTURAL COLLEGE, Agricultural College,
[1855], 1857

△EXPERIMENT STATION OF MICHIGAN AGRICULTURAL COL-
LEGE, Agricultural College, 1888

Historical Society of Michigan, Detroit

No academy of sciences

MINNESOTA

(Territory, 1849; State, 1858; land grants, 1857, 1861, and 1870)
†*UNIVERSITY OF MINNESOTA, Minneapolis [1857], 1868
COLLEGE OF AGRICULTURE AND MECHANIC ARTS OF THE UNI-
VERSITY OF MINNESOTA, Saint Anthony Park, 1868
†STATE SCHOOL OF AGRICULTURE OF THE UNIVERSITY OF
MINNESOTA, Saint Anthony Park, 1888
△AGRICULTURAL EXPERIMENT STATION OF THE UNIVERSITY
OF MINNESOTA, Saint Anthony Park, 1888
Minnesota Historical Society, St. Paul
Minnesota Academy of Science, Minneapolis, 1873
St. Paul Academy of Sciences, St. Paul

MISSISSIPPI

(Territory, 1798, State, 1817; land grants 1803, 1819)
[JEFFERSON COLLEGE, Washington, 1803–discontinued]
†UNIVERSITY OF MISSISSIPPI, Oxford, 1874
*AGRICULTURAL AND MECHANICAL COLLEGE OF MISSISSIPPI,
Agricultural College (Starkville), 1880
△MISSISSIPPI AGRICULTURAL EXPERIMENT STATION, Agricultural
College, 1888
*ALCORN AGRICULTURAL AND MECHANICAL COLLEGE (colored),
Rodney, 1871, reorganized in 1878
Mississippi Historical Society, Jackson
No academy of sciences

MISSOURI

(Territory, 1812; State, 1821; land grants, 1818 and 1820)
†*UNIVERSITY OF MISSOURI, Columbia [1820], 1839
MISSOURI AGRICULTURAL AND MECHANICAL COLLEGE OF THE
UNIVERSITY OF MISSOURI, Columbia, 1870
△MISSOURI AGRICULTURAL EXPERIMENT STATION, Columbia,
1881

*MISSOURI SCHOOL OF MINES AND METALLURGY OF THE UNIVER-
SITY OF MISSOURI, Rolla, 1870
Missouri Historical Society, St. Louis
St. Louis Academy of Sciences, 1857

MONTANA
(Territory, 1864; land grant, 1881)
COLLEGE OF MONTANA, Deer Lodge, 1883
Montana Historical Society, Helena

NEBRASKA
(Territory, 1859; State, 1867; land grant, 1881)
†*UNIVERSITY OF NEBRASKA, Lincoln, 1869
INDUSTRIAL COLLEGE OF THE UNIVERSITY OF NEBRASKA, Lin-
coln, 1869; opened for students 1871
△AGRICULTURAL EXPERIMENT STATION OF NEBRASKA, Lincoln,
1887
Nebraska State Historical Society, Lincoln, 1878
No scientific society

NEVADA
(Territory, 1861; State, 1864; land grant, 1866)
†*STATE UNIVERSITY OF NEVADA, Reno [1865], 1874
SCHOOL OF AGRICULTURE OF THE NEVADA STATE UNIVERSITY,
Reno, 1877
△NEVADA STATE AGRICULTURAL STATION, Reno
No scientific or historical society

NEW HAMPSHIRE
(Settled, 1629; State, 1788)
DARTMOUTH COLLEGE, Hanover [1758], 1770
*NEW HAMPSHIRE COLLEGE OF AGRICULTURE AND THE ME-
CHANIC ARTS (in connection with Dartmouth College), Hanover
[1866], 1868

△NEW HAMPSHIRE AGRICULTURAL EXPERIMENT STATION, Hano-
　　ver, 1888
New Hampshire Historical Society, Concord, 1823
No academy of science

NEW JERSEY
(Settled, 1614–20; State, 1787)
COLLEGE OF NEW JERSEY, Princeton, 1746
*RUTGERS SCIENTIFIC SCHOOL OF RUTGERS COLLEGE, New Bruns-
　　wick. Made State College of Agriculture and the Mechanic Arts
　　[1864], 1865
†NEW JERSEY STATE AGRICULTURAL EXPERIMENT STATION,
　　New Brunswick, 1880
△NEW JERSEY AGRICULTURAL COLLEGE EXPERIMENT STATION,
　　New Brunswick, 1888
New Jersey Historical Society, Newark, 1845
No academy of science

NEW MEXICO
(Territory, 1850; land grant, 1854)
UNIVERSITY OF NEW MEXICO, Santa Fe, 1881
†AGRICULTURAL COLLEGE OF NEW MEXICO, Las Cruces. Established
　　by Territorial legislature, 1888–89
Historical Society of New Mexico, Santa Fe

NEW YORK
(Settled 1613; State, 1788)
THE UNIVERSITY OF NEW YORK, 1787, is not a teaching body. It is in
　　indirect relationship with Columbia College, 1754, Union College,
　　Hamilton College, and numerous collegiate and technical schools.
*CORNELL UNIVERSITY, Ithaca [1865], 1868
COLLEGE OF AGRICULTURE OF CORNELL UNIVERSITY, Ithaca, 1888
△CORNELL UNIVERSITY AGRICULTURAL EXPERIMENT STATION,
　　Ithaca, 1879

†NEW YORK AGRICULTURAL EXPERIMENT STATION, Geneva, 1882
New York Historical Society, New York, 1804
New York Academy of Sciences, 1817

NORTH CAROLINA
(Settled, 1653; State, 1789)
†UNIVERSITY OF NORTH CAROLINA, Chapel Hill [1789], 1795
*NORTH CAROLINA COLLEGE OF AGRICULTURE AND MECHANIC
ARTS, Raleigh. Established by State, 1889
△†NORTH CAROLINA AGRICULTURAL EXPERIMENT STATION, Ra-
leigh, 1877 and 1887

OHIO
(Territory, 1788; States, 1803; land grants, 1792 and 1803)
OHIO UNIVERSITY, Athens, 1804
MIAMI UNIVERSITY, Oxford, 1809, 1816
†*OHIO STATE UNIVERSITY, Columbus. Chartered 1870; organized Septem-
ber 17, 1873
△OHIO AGRICULTURAL EXPERIMENT STATION, Columbus, 1882
and 1888
Historical and Philosophical Society of Ohio, Cincinnati
No State scientific society

OREGON
(Territory, 1848; State, 1859)
†UNIVERSITY OF OREGON, Eugene City [1850], 1876
*OREGON STATE AGRICULTURAL COLLEGE, Corvallis, 1888
△OREGON EXPERIMENT STATION, Corvallis, 1888
Pioneer and Historical Society, Astoria
No scientific society

PENNSYLVANIA
(Settled, 1626; State, 1787)
UNIVERSITY OF PENNSYLVANIA, Philadelphia, 1751

*PENNSYLVANIA STATE COLLEGE, State College, 1859, 1862, and 1874
   △†PENNSYLVANIA STATE COLLEGE AGRICULTURAL EXPERIMENT
      STATION, State College, 1887
   American Philosophical Society, Philadelphia, 1769
   Historical Society of Pennsylvania, Philadelphia, 1824

RHODE ISLAND
(Settled 1636; State, 1790)

*BROWN UNIVERSITY, Providence, 1764
AGRICULTURAL AND SCIENTIFIC DEPARTMENT OF BROWN
   UNIVERSITY, Providence
†RHODE ISLAND STATE AGRICULTURAL SCHOOL, Kingston, 1888
   △RHODE ISLAND STATE AGRICULTURAL COLLEGE EXPERIMEN-
      TAL STATION, Kingston, 1888
   Rhode Island Historical Society, Providence

SOUTH CAROLINA
(Settled, 1670; State, 1788)

*†UNIVERSITY OF SOUTH CAROLINA, Columbia, 1801; reorganized,
   1865
SOUTH CAROLINA COLLEGE OF AGRICULTURE AND MECHANIC
   ARTS, UNIVERSITY OF SOUTH CAROLINA, Columbia 1879
   △SOUTH CAROLINA AGRICULTURAL EXPERIMENT STATION,
      Columbia, 1888
*CLAFLIN UNIVERSITY AND SOUTH CAROLINA AGRICULTURAL
   COLLEGE AND MECHANICS' INSTITUTE (Department of Uni-
   versity of South Carolina), Orangeburg, 1872
   South Carolina Historical Society, Charleston

TENNESSEE
(Territory, 1790; State, 1796)

UNIVERSITY OF NASHVILLE (Cumberland College), 1806;
   discontinued 1875
†*UNIVERSITY OF TENNESSEE, Knoxville, 1806

STATE AGRICULTURAL AND MECHANICAL COLLEGE OF THE UNI-
VERSITY OF TENNESSEE, Knoxville, 1869
△TENNESSEE AGRICULTURAL EXPERIMENT STATION, Knoxville,
1882 and 1887
Tennessee Historical Society, Nashville

TEXAS
(Annexed, 1846; State, 1845)
†UNIVERSITY OF TEXAS, Austin [1839], 1866
*STATE AGRICULTURAL AND MECHANICAL COLLEGE OF TEXAS, Col-
lege Station [1871], 1876
△TEXAS AGRICULTURAL EXPERIMENT STATION, College Station,
1888
No historical or scientific society

UTAH
(Territory, 1850; land grant, 1855)
UNIVERSITY OF DESERET, Salt Lake City, 1850
†UTAH AGRICULTURAL COLLEGE, Logan City. Established by Territorial
legislature, March 8, 1888

VERMONT
(Settled, 1755–58; State, 1791)
*UNIVERSITY OF VERMONT [1791], 1800, and
STATE AGRICULTURAL COLLEGE, Burlington, 1865–67
△†VERMONT STATE AGRICULTURAL EXPERIMENT STATION, Burling-
ton, 1887
Vermont Historical Society, Montpelier

VIRGINIA
(Settled, 1609; State, 1788)
[COLLEGE OF HENRICO. Projected in 1620]
WILLIAM AND MARY COLLEGE, Williamsburgh, 1691
†UNIVERSITY OF VIRGINIA, Charlottesville, 1819

*VIRGINIA AGRICULTURAL AND MECHANICAL COLLEGE, Blacksburg,
    1872
△VIRGINIA AGRICULTURAL EXPERIMENT STATION, Blacksburg, 1888
*HAMPTON NORMAL AND AGRICULTURAL INSTITUTE, Hampton.
    Organized by American Missionary Society, April, 1868;
    reorganized under charter from State, June, 1870
    Virginia Historical Society, Richmond, 1831

WASHINGTON
(Territory, 1853; State, 1889)
UNIVERSITY OF WASHINGTON, Seattle, 1862

WEST VIRGINIA
(State, 1862)
†*WEST VIRGINIA UNIVERSITY, Morgantown, 1867
AGRICULTURAL DEPARTMENT OF WEST VIRGINIA UNIVERSITY, Mor-
    gantown
△WEST VIRGINIA EXPERIMENT STATION, Morgantown, 1888
    West Virginia Historical Society, Morgantown

WISCONSIN
(Territory, 1836; State, 1847; land grants, 1846 and 1854)
†*UNIVERSITY OF WISCONSIN, Madison [1838], 1848
DEPARTMENT OF AGRICULTURE OF THE UNIVERSITY OF
    WISCONSIN, Madison, 1866
△†AGRICULTURAL EXPERIMENT STATION OF THE UNIVERSITY OF
    WISCONSIN, Madison, 1883 and 1888
    Wisconsin Historical Society, Madison
    Wisconsin Academy of Science, Arts, and Letters, Madison, 1870

WYOMING
(Territory, 1868; State, 1889)
UNIVERSITY OF WYOMING, Laramie City
    Wyoming Academy of Arts, Science, and Letters, Cheyenne

MUSEUM-HISTORY AND MUSEUMS OF HISTORY[1]

By George Brown Goode

Assistant Secretary, Smithsonian Institution, in charge of the U.S. National Museum

The true significance of the word museum may perhaps best be brought to our apprehension by an allusion to the ages which preceded its origin—when our ancestors, hundreds of generations removed, were in the midst of those great migrations which peopled Europe with races originally seated farther to the east.

It has been well said that the story of early Greece is the first chapter in the history of the political and intellectual life of Europe.

To the history of Greece let us go for the origin of the museum idea, which in its present form seems to have found its only congenial home among the European offshoots of the Indo-Germanic division of the world's inhabitants.

Museums, in the language of ancient Greece, were the homes of the muses. The first were in the groves of Parnassus and Helicon, and later they were temples in various parts of Hellas. Soon, however, the meaning of the word changed, and it was used to describe a place of study, or a school. Athenæus described Athens in the second century as "the museum of Greece," and the name of museum was definitely applied to that portion of the palace of Alexandria which was set apart for the study of the sciences, and which contained the famous Alexandrian library. The museum of Alexandria was a great university, the abiding place of men of science and letters, who were divided into many companies or colleges, and for whose support a handsome revenue was allotted.

The Alexandrian museum was destroyed in the days of Caesar and Aurelian, and the term museum, as applied to a great public institution, dropped out of use from the fourth to the seventeenth century. The disappearance of a word is an indication that the idea for which it stood has also fallen into disfavor; and such, indeed, was the fact. The history of museum and library run in parallel lines. It is not until the development of the arts and sciences has taken place, until an extensive written literature has grown up, and a distinct literary and scientific class has been developed, that it is possible for the modern library and museum to come into existence. The museum of the present is more dissimilar to its old-time representative than is our library to its prototype.

There were in the remote past galleries of pictures and sculpture, as well as so-called museums. Public collections of paintings and statuary were founded in Greece and Rome at a very early day. There was a gallery of paintings (Pinacotheca) in one of the marble halls of the propylæum at Athens, and in Rome there were lavish public displays of works of art. M. Dezobry, in his Rome in the Time of Augustus, has described this phase of Latin civilization in the first century before Christ:

For many years [remarks one of his characters] the taste for paintings has been extending in a most extraordinary manner. In former times they were only to be found in the temples, where they were placed less for purposes of ornament than as an act of homage to the gods; now they are everywhere, not only in temples, in private houses, and in public halls, but also on outside walls, exposed freely to air and sunlight. Rome is one great picture gallery; the

Forum of Augustus is gorgeous with paintings, and they may be seen also in the Forum of Cæsar, in the Roman Forum, under the peristyles of many of the temples, and especially in the porticos used for public promenades, some of which are literally filled with them. Thus everybody is enabled to enjoy them, and to enjoy them at all hours of the day.

The public men of Rome, at a later period in its history, were no less mindful of the claims of art. They believed that the metropolis of a great nation should be adorned with all the best products of civilization. We are told by Pliny that when Cæsar was dictator, he purchased, for 300,000 deniers, two Greek paintings, which he caused to be publicly displayed, and that Agrippa placed many costly works of art in a hall which he built and bequeathed to the Roman people. Constantine gathered together in Constaninople the paintings and sculptures of the great masters, so that the city, before its destruction, became a great museum, like Rome.

The taste for works of art was generally prevalent throughout the whole Mediterranean region in the days of the ancient civilizations, and there is abundant reason to believe that there were prototypes of the modern museum in Persia, Assyria, Babylonia, and Egypt, as well as in Rome. Collections in natural history also undoubtedly existed, though we have no positive descriptions of them. Natural curiosities, of course, found their way into the private collections of monarchs, and were doubtless also in use for study among the savants in the Alexandrian museum. Aristotle, in the fourth century before Christ, had, it is said, an enormous grant of money for use in his scientific researches, and Alexander the Great, his patron, "took care to send to him a great variety of zoological specimens, collected in the countries which he had subdued," and also "placed at his disposal several thousand persons, who were occupied in hunting, fishing, and making the observations which were necessary for completing his History of Animals." If human nature has not changed more than we suppose, Aristotle must have had a great museum of natural history.

When the Roman capital was removed to Byzantium, the arts and letters of Europe began to decline. The Church was unpropitious, and the invasions of the northern barbarians destroyed everything. In 476, with the close of the Western Empire, began a period of intellectual torpidity which was to last for a thousand years.

It was in Bagdad and Cordova that science and letters were next to be revived, and Africa was to surpass Europe in the extent of its libraries. In the Periplus, or Voyage of Hanno, occurs the following passage in regard to specimens of Gorillas, or "Gorgones":

Pursuing them, we were not able to take the men (males); they all escaped, being able to climb the precipices, and defended themselves with pieces of rock. But three women (females), who bit and scratched those who led them, were not willing to follow. However, having killed them, we flayed them, and conveyed the skins to Carthage; for we did not sail any further, as provisions began to fail.[2]

With the Renaissance came a period of new life for collectors. The churches of southern Europe became art galleries, and monarchs and noblemen and ecclesiastical dignitaries collected books, manuscripts, sculptures, pottery, and gems, forming the beginning of collections which have since grown into public museums. Some of these collections doubtless had their first beginnings in the midst of the dark ages, within the walls of feudal castles, or the larger monasteries, but their number was small, and they must have consisted chiefly of those objects so nearly akin to literature as especially to command the attention of bookish men.

As soon as it became the fashion for the powerful and the wealthy to possess collections, the scope of their collections began to extend, and objects were gathered on account of their rarity or grotesqueness, as well as for their beauty or instructiveness. Flourens, in his Life and Works of Blumenbach, remarks: "The old Germany, with its old chateaux, seemed to pay no homage to science; still the lords of these ancient and noble mansions had long since made it a business, and almost a point of honor, to form with care what were called cabinets of curiosities."

To the apothecary of old, with his shop crowded with the curious substances used in the medical practice of his day, the museum owes some of its elements, just as the modern botanic garden owes its earliest history to the "physic garden," which in its time was an outgrowth of the apothecary's garden of simples. The apothecary in Romeo and Juliet—

> In whose needy shop a tortoise hung,
> An alligator stuff'd, and other skins
> Of eel-shaped fishes,—

was the precursor of the modern museum keeper. In the hostelries and taverns, the gathering places of the people in the sixteenth and seventeenth centuries, there grew up little museums of curiosities from foreign lands, while in the great fairs were always exhibited sundry gatherings of strange and entertaining objects.

At the middle of the last century there appear to have been several such collections of curiosities in Britain.

In Artedi's ichthyological works there are numerous references to places where he had seen American fishes, especially at Spring Garden (later known as the Vauxhall Garden, a famous place of resort), and at the Nag's head, and the White Bear, and the Green Dragon in Stepney, in those days a famous hostelry in London. He speaks also of collections at the houses of Mr. Lillia and in that of Master Saltero (the barber-virtuoso, described by Bulwer in his Devereux), in Chelsea and at Stratford, and also in the collection of Seba, in Amsterdam, and in that of Hans Sloane.

With the exception of "*the monk* or *Angel-fish*, *Anglis*, alias *Mermaid-fish*," probably a species of *Squatina*, which he saw at the Nag's Head, all the fishes in these London collections belonged to the order Plectognathi.

Josselyn, in his *Two Voyages to New England* (1638--1673), after telling us how a Piscataway colonist had the fortune to kill a Pilhannaw—the king of the birds of prey—continues, "How he disposed of her I know not, but had he taken her alive and sent her over into England, neither Bartholomew or Sturbridge Fair could have produced such another sight."

Shakespeare's mirror strongly reflects the spirit of the day. When Trinculo, cast ashore upon a lonesome island, catches a glimpse of Caliban, he exclaims:

What have we here? A man or a fish? Dead or alive? A fish: he smells like a fish; a very ancient and fish-like smell. . . . a strange fish! Were I in England now (as once I was), and had but this fish painted, not a holiday fool there but would give a piece of silver; there would this monster make a man; any strange beast there makes a man: when they will not give a doit to relieve a lame beggar, they will lay out ten to see a dead Indian.

The idea of a great national museum of science and art was first worked out by Lord Bacon in his *New Atlantis*, a philosophical romance published at the close of the seventeenth century. The first scientific museum actually founded was that begun at Oxford in 1667, by Elias Ashmole, still known as the Ashmolean Museum, composed chiefly of natural history specimens collected by the botanists Tradescant, father and son, in Virginia, and in the north of Africa. Soon after, in 1753, the British Museum was established by act of Parliament, inspired by the will of Sir Hans Sloane, who, dying in 1749, left to the nation his invaluable collection of books, manuscripts, and curiosities.[3]

Many of the great national museums of Europe had their origin in the private collections of monarchs. France claims the honor of having been the first to change a royal into a national museum, when, in 1789, the Louvre came into the possession of a republican government. It is very clear, however, that democratic England, by its action in 1753, stands several decades in advance—its act, moreover, being one of deliberate founding rather than a species of conquest.

The first chapter in the history of American museums is short. In colonial days there were none. In the early years of the Republic, the establishment of such institutions by city, State, or Federal Government would not have been considered a legitimate act. When the General Government came into the possession of extensive collections as the result of the Wilkes Exploring Expedition in 1842, they were placed in charge of a private organization, the National Institution, and later, together with other similar materials, in that of a corporation, the Smithsonian Institution, which was for a long period of years obliged to pay largely for their care out of its income from a private endowment. It was not until 1876, however, that the existence of a National Museum, as such, was definitely recognized in the proceedings of Congress, and its financial support fully provided for.

In early days, however, our principal cities had each a public museum, founded and supported by private enterprise. The earliest general collection was that formed at Norwalk, Connecticut, prior to the Revolution, by a man named Arnold, described as "a curious collection of American birds and insects." This it was which first awakened the interest of President Adams in the natural sciences. He visited it several times as he traveled from Boston to Philadelphia, and his interest culminated in the foundation of the American Academy of Arts and Sciences.[4] In 1790 Doctor Hosack brought to America from Europe the first cabinet of minerals ever seen on this continent.

The earliest public establishment, however, was the Philadelphia Museum, established by Charles Willson Peale in 1785, which had for a nucleus a stuffed paddlefish and the bones of a mammoth, and which was for a time housed in the building of the American Philosophical Society. In 1800 it was full of popular attractions.

There were a mammoth's tooth from the Ohio, and a woman's shoe from Canton; nests of the kind used to make soup of, and a Chinese fan six feet long; bits of asbestus, belts of wampum, stuffed birds and feathers from the Friendly Islands, scalps, tomahawks, and long lines of portraits of great men of the Revolutionary war. To visit the museum, to wander through the rooms, play upon the organ, examine the rude electrical machine, and have a profile drawn by the physiognomitian, were pleasures from which no stranger to the city ever refrained.

Doctor Hare's oxyhydrogen blowpipe was shown in this museum by Mr. Rubens Peale as early as 1810.

The Baltimore Museum was managed by Rembrandt Peale, and was in existence as early as 1815 and as late as 1830.

Earlier efforts were made, however, in Philadelphia. Doctor Chovet, of that city, has a collection of wax anatomical models made by him in Europe, and Professor John Morgan, of the University of Pennsylvania, who learned his methods from the Hunters in London and Sué in Paris, was also forming such a collection before the Revolution.

The Columbian Museum and Turrel's Museum, in Boston, are spoken of in the annals of the day, and there was a small collection in the attic of the statehouse in Hartford.

The Western Museum, in Cincinnati, was founded about 1815 by Robert Best, M. D., afterwards of Lexington, Kentucky, who seems to have been a capable collector, and who contributed matter to Godman's American Natural History. In 1818 a society styled the Western Museum Society was organized among the citizens, which, though scarcely a scientific organization, seems to have taken a somewhat liberal and public-spirited view of what a museum should be. With the establishment of the Academy of Natural Sciences in Philadelphia in 1812, and the New York Lyceum of Natural History, the history of American scientific museums had its true beginning.

The intellectual life of America is so closely allied to that of England that the revival of interest in museums and in popular education at the middle of the present century is especially significant to us. The great exhibition of 1851 was one of the most striking features of the industrial revolution in England, that great transformation which, following closely upon the introduction of railroads, turned England feudal and agricultural into England democratic and commercial.

The great exhibition marked an epoch in the intellectual progress of English-speaking people. "The great exhibition," writes a popular novelist, and a social philosopher as well, "did one great service for country people. It taught them how easy it is to get to London, and what a mine of wealth, especially for after memory and purposes of conversation, exists in that great place.

Under the wise administration of the South Kensington staff, a great system of educational museums has been developed all through the United Kingdom.

Our own Centennial Exhibition in 1876 was almost as great a revelation to the people of the United States. The thoughts of the country were opened to many things before undreamed of. One thing we may regret—that we have no such widespread system of museums as that which has developed in the motherland with South Kensington as its administrative center. England has had nearly forty years, however, and we but thirteen, since our exhibition. May we not hope that within a like period of time, and before the year 1914, the United States may have attained the position which England now occupies, at least in the respects of popular interest and substantial governmental support? There are now over one hundred and fifty public museums in the United Kingdom, all active and useful.

The museum systems of Great Britain are, it seems to me, much closer to the ideal which America should follow, than are those of either France or Germany. They are designed more thoughtfully to meet the needs of the people, and are more intimately intertwined with the policy of national popular education.

Sir Henry Cole, the working founder of the Department of Science and Art, speaking of the purpose of the museums under his care, said to the people of Birmingham in 1874:

If you wish your schools of science and art to be effective, your health, the air, and your food to be wholesome, your life to be long, your manufactures to improve, your trade to increase, and your people to be civilized, you must have museums of science and art to illustrate the principles of life, health, nature, science, art, and beauty.

Again, in words as applicable to Americans of to-day as to Britons in 1874, said he:

A thorough education and a knowledge of science and art are vital to the nation, and to the place it holds at present in the civilized world. Science and art are the lifeblood of successful production. All civilized nations are running a race with us, and our national decline will date from the period when we go to sleep over the work of education, science, and art. What has been done is at the mere threshold of the work yet to be done.

The people's museum should be much more than a house full of specimens in glass cases. It should be a house full of ideas, arranged with the strictest attention to system. I once tried to express this thought by saying: "An efficient educational museum may be described as a collection of instructive labels, each illustrated by a well-selected specimen."

The museum, let me add, should be more than a collection of specimens, well arranged and well labeled. Like the library, it should be under the constant supervision of one or more men, well informed, scholarly, and withal practical, and fitted by tastes and training to aid in the educational work. I should not organize the museums primarily for the use of people in their larval or school-going stage of existence. The public school-teacher, with the illustrated text-books, diagrams, and other appliances, has in these days a professional outfit which is usually quite sufficient to enable him to teach his pupils.

School days last at the most only from four to fifteen years, and they end, with the majority of mankind, before their minds have reached the stage of growth most favorable for the reception and assimilation of the best and most useful thought. Why should we be crammed in the time of infancy and kept in a state of mental starvation during the period which follows, from maturity to old age—a state which is disheartening and unnatural all the more because of the intellectual tastes which have been stimulated and partially formed by school life?

The museum idea is much broader than it was fifty or even twenty-five years ago. The museum of to-day is no longer a chance assemblage of curiosities, but rather a series of objects selected with reference to their value to investigators, or their possibilities for public enlightenment. The museum of the future may be made one of the chief agencies of the higher civilization.

I hope that the time will come when every town shall have both its public museum and its public library, each with a staff of competent men, mutually helpful, and contributing largely to the intellectual life of the community.

The museum of the future in this democratic land should be adapted to the needs of the mechanic, the factory operator, the day laborer, the salesman, and the clerk, as much as to those of the professional man and the man of leisure. It is proper that there be laboratories and professional libraries for the development of the experts who are to organize, arrange, and explain the museums.

It is proper that laboratories be utilized to the fullest extent for the credit of the institution to which they belong. No museum can do good and be respected which does not each year give additional proofs of its claims to be considered a center of learning. On the other hand, the public have a right to ask that much shall be done directly in their interest. They will gladly allow the museum officer to use part of his time in study and experiment. They will take pride in the possession by the museum of tens of thousands of specimens, interesting only to the specialist, hidden away perpetually from public view, but necessary for proper scientific research. They are the foundations of the intellectual superstructure which gives to the institution its proper standing.

Still, no pains must be spared in the presentation of the material in the exhibition halls. The specimens must be prepared in the most careful and artistic manner, and arranged attractively in well-designed cases and behind the clearest of glass. Each object must bear a label giving its name and history so fully that all the probable questions of the visitor are answered in advance. Books of reference must be kept in convenient places. Colors of walls, cases, and labels must be restful and quiet, and comfortable seats must be everywhere accessible, for the task of the museum visitor is a weary one at best.

All intellectual work may be divided into two classes, the one tending toward the increase of knowledge, the other toward its diffusion; the one toward investigation and discovery, the other toward the education of the people and the application of known facts to promoting their material welfare. The efforts of learned men and of institutions of learning are sometimes applied solely to one of these departments of effort—sometimes to both—and it is generally admitted, by the most advanced teachers, that, for their students as well as for themselves, the happiest results are reached by carrying on investigation and instruction simultaneously. Still more is this true of institutions of learning. The college which imparts only second-hand knowledge to its students belongs to a period in the history of education which is fast being left behind.

The museum must, in order to perform its proper functions, contribute to the advancement of learning through the increase as well as through the diffusion of knowledge.

We speak of "educational" museums and of the "educational" method of installation so frequently that there may be danger of inconsistency in the use of the term. An educational museum, as it is usually spoken of, is one in which an attempt is made to teach the unprofessional visitor of an institution for popular education by means of labeled collections, and it may be, also, by popular lectures. A college museum, although used as an aid to advanced instruction, is not an "educational museum" in the ordinary sense, nor does a museum of research, like the Museum of Comparative Zoology at Cambridge, Massachusetts, belong to this class, although, to a limited extent, it attempts and performs popular educational work in addition to its other functions.

In the National Museum in Washington the collections are divided into two great classes. The exhibition series, which constitutes the educational portion of the Museum, and is exposed to public view, with all possible accessories for public entertainment and instruction; and the study series, which is kept in the scientific laboratories, and is rarely examined except by professional investigators.

In every properly conducted museum the collections must, from the very beginning, divide themselves into these two classes, and, in planning for its administration, provision should be made not only for the exhibition of objects in glass cases, but for the preservation of large collections not available for exhibition, to be used for the studies of a very limited number of specialists. Lord Bacon, who, as we have noticed, was the first to whom occurred the idea of a great museum of science and art, complains thus, centuries ago, in his book, On the Advancement of Learning, that up to that time the means for intellectual progress had been used exclusively for "amusement" and "teaching," and not for the "augmentation of science."

The boundary line between the library and the museum is neither straight nor plain. The former, if its scope be rightly indicated by its name, is, primarily, a place for books. The latter is a depository for objects of every kind, books not excepted. The British Museum, with its libraries, its pictures, its archæological galleries, its anthropological, geological, botanical, and zoological collections, is an example of the most comprehensive interpretation of the term. Professor Huxley has described the museum as "a consultative library of objects." This definition is suggestive but unsatisfactory. It relates only to the contents of the museum as distinguished from those of the library, and makes no reference to the differences in the methods of their administration.

The treasures of the library must be examined one at a time, and by one person at a time. Their use requires long-continued attention, and their removal from their proper places in the system of arrangement. Those of the museums are displayed to public view in groups, in systematic sequence, so that they have a collective as well as an individual significance. Furthermore, much of their meaning may be read at a glance. The museum cultivates the powers of observation, and the casual visitor even makes discoveries for himself, and, under the guidance of the labels, forms his own impression. In the library one studies the impressions of others.

The library is most useful to the educated; the museum to educated and uneducated alike, to the masses as well as to the few, and is a powerful stimulant to intellectual activity in either class.

The influence of the museum upon a community is not so deep as that of the library, but extends to a much larger number of people. The National Museum in Washington has 300,000 visitors a year, each of whom carries away a certain number of new thoughts.

The two ideas may be carried out, side by side, in the same building, and, if need be, under the same management, not only without antagonism, but with advantage. That the proximity of a good library is absolutely essential to the influence of a museum, will be admitted by every-one. I am confident, also, that a museum wisely organized and properly arranged is certain to benefit the library near which it stands in many ways, and more positively than through its power to stimulate interest in books, and thus to increase the general popularity of the library and to enlarge its endowment.

Many books and valuable ones would be required in this best kind of museum work, but it is not intended to enter into competition with the library. When necessary, volumes might be duplicated. It is very often the case, however, that books are more useful and safer in the museum than on the library shelves, for in the museum they may be seen daily by thousands, while in the library their very existence is forgotten by all except their custodian.

Audubon's Birds of North America is a book which everyone has heard of and which every one wants to see at least once in his lifetime. In a library, it probably is not examined by ten persons in a year. In a museum, if the volume were exposed to view in a glass case, a few of the most striking plates detached, framed, and hung upon the wall near at hand, it will teach a lesson to every passer-by.

The library may be called upon for aid by the museum in many direc-tions. Pictures are often better than specimens to illustrate certain ideas. The races of man and their distribution can only be shown by pictures and maps. Atlases of ethnological portraits and maps are out of place in a library if there is a museum nearby in which they can be displayed. They are not even members of the class described by Lamb as "books which are not books." They are not books, but museum specimens, mas-querading in the dress of books.

In selecting courses for the development of a museum, it may be useful to consider what are the fields open to museum work. As a matter of convenience, museums are commonly classed in two groups—those of science and those of art—and in Great Britain the great national system is mainly under the control of The Science and Art Department of the Committee of Council on Education.

This classification is not entirely satisfactory, since it is based upon methods of arrangement rather than upon the nature of the objects to be arranged, and since it leaves in a middle territory (only partially occupied by the English museum men of either department) a great mass of museum material, of the greatest moment, both in regard to its interest and its adaptability for purposes of public instruction.

On the other side stand the natural history collections, undoubtedly best to be administered by the geologist, botanist, and zoologist. On the other side are the fine-art collections, best to be arranged, from an æsthetic standpoint, by artists. Between is a territory which no English word can adequately describe—which the Germans call *Culturgeschichte*—the natural history of civilization, of man and his ideas and achievements. The museums of science and art have not yet learned how to partition this territory.

An exact classification of museums is not at present practicable, nor will it be until there has been some redistribution of the collections which they contain. It may be instructive, however, to pass in review the principal museums of the world, indicating briefly their chief characteristics.

Every great nation has its museum of natural history. The natural history department of the British Museum, recently removed from the heart of London to palatial quarters in South Kensington, is probably the most extensive, with its three great divisions, zoological, botanical, and geological.

The historian and the naturalist have met upon common ground in the field of anthropology. The anthropologist is, in most cases, historian as well as naturalist; while the historian of to-day is always in some degree an anthropologist, and makes use of many of the methods at one time peculiar to the natural sciences. The museum is no less essential to the study of anthropology than to that of natural history. The library formerly afforded to the historian all necessary opportunities for work. It would seem from the wording of the new charter of the American Historical Association that its members consider a museum to be one of its legitimate agencies.

Your secretary has invited me to say something about the possibilities of utilizing museum methods for the promotion of historical studies. This I do with much hesitation, and I hope that my remarks may be considered as suggestions rather than as expressions of definite opinion. The art of museum administration is still in its infancy, and no attempt has yet been made to apply it systematically to the development of a museum of history. Experiment is as yet the museum administrator's only guide, and he often finds his most cherished plans thoroughly impracticable. That museums can ever be made as useful to history as they are to physical science, their most enthusiastic friend dares not hope. The two departments of science are too unlike.

The historian studies events and their causes; the naturalist studies objects and the forces by which their existence is determined. The naturalist may assemble in a museum objects from every quarter of the globe and from every period of the earth's history. Much of his work is devoted to the observation of finished structure, and for this purpose his specimens are at all times ready. When, however, he finds it necessary to study his subject in other aspects, he may have recourse to the physical, chemical, and physiological laboratories, the zoological and botanical gardens, and aquaria, which should form a part of every perfect museum system. Here, almost at will, the phenomena of nature may be scrutinized and confirmed by repeated observation, while studies impracticable in the nursery may usually be made by members of its staff, who carry its appliances with them to the seashore or to distant lands.

The requirements of the historian are very different. Nevertheless, I am confident that the museum may be made in his hands a most potent instrumentality for the promotion of historical studies. Its value is perhaps less fully realized than it would be were it not that so many of its functions are performed by the library. In the library may be found descriptive catalogues of all the great museums, and books by the hundred, copiously illustrated with pictures of the objects preserved in museums. A person trained to use books may by their aid reap the advantage of many museums without the necessity of a visit to one.

The exhibition series would be proportionately larger in an historical than in a natural-history museum. The study series of a historical museum would mostly be arranged in the form of a library, except in some special departments, such as numismatics, and when a library is near might be entirely dispensed with.

The adoption of museum methods would be of advantage to the historian in still another way, by encouraging the preservation of historical material not at present sought for by librarians, and by inducing present owners of such material to place it on exhibition in public museums.

Although there is not in existence a general museum of history arranged on the comprehensive plan adopted by natural-history museums, there are still many historical collections of limited scope, which are all that could be asked, and more.

The value to the historian of archæological collections, historic and prehistoric, has long been understood. The museums of London, Paris, Berlin and Rome need no comment. In Cambridge, New York, and Washington are immense collections of the remains of man in America in the pre-Columbian period—collections which are yearly growing in significance, as they are made the subject of investigation, and there is an immense amount of material of this kind in the hands of institutions and private collectors in all parts of the United States.

The museum at Naples shows, so far as a museum can, the history of Pompeii at one period. The museum of St. Germain, near Paris, exhibits the history of France in the time of the Gauls and of the Roman occupation. In Switzerland, especially at Neuchatel, the history of the inhabitants of the Lake Dwellings is shown.

American ethnological museums are preserving with care the memorials of the vanishing race of red men. The George Catlin Indian Gallery, which is installed in the room in which this society is now meeting, is valuable beyond the possibility of appraisement, in that it is the sole record of the physical characters, the costumes, and the ceremonies of several tribes long extinct.

Other countries recently settled by Europeans are preserving the memorials of the aboriginal races, notably the colonies in Australia and New Zealand. Japan is striving to preserve in its Government museum examples of the fast-disappearing memorials of feudal days.

Ethnographic museums are especially numerous and fine in the northern part of Europe. They were proposed more than half a century ago, by the French geographer, Jomard, and the idea was first carried into effect about 1840, on the establishment of the Danish Ethnographical Museum, which long remained the best in Europe. Within the past twenty years there is an extraordinary activity in this direction.

In Germany, besides the chief museum in Berlin, considerable ethnographical collections have been founded in Hamburg and Munich. Austria has in Vienna two for ethnography, the Court Museum (Hof-Museum), and the Oriental (Orientalisches Museum). Holland has reorganized the National Ethnographical Museum (Rÿks Ethnographisch Museum) in Leyden, and there are smaller collections in Amsterdam, Rotterdam, and The Hague. France has founded the Trocadero (Musée de Trocadero). In Italy there is the important Prehistoric and Ethnographic Museum (Museo preistorico ed ethnografico) in Rome, as well as the collection of the Propaganda, and there are museums in Florence and Venice.

Ethnographical museums have also been founded in Christiania and Stockholm, the latter of which will include the rich material collection by Doctor Stolpe on the voyage of the frigate *Vanadis* around the world.

In England there is less attention to the subject, the Christy collection in the British Museum being the only one specially devoted to ethnography, unless we include also the local Blackmore Museum at Salisbury.

In the United States the principal establishment arranged on the ethnographic plan is the Peabody Museum of Archæology in Cambridge, and there are important smaller collections in the American Museum of Natural History in New York and the Peabody Academy of Sciences at Salem.

The ethnological collections in Washington are classified on a double system, in one of its features corresponding to that of the European, in the other like the famous Pitt-Rivers collection at Oxford, arranged to show the evolution of culture and civilization without regard to race. This broader plan admits much material excluded by the advocates of ethnographic museums, who devote their attention almost exclusively to the primitive or non-European peoples.

In close relation to the ethnographic museums are those which are devoted to some special field of human thought and interest. Most remarkable among these probably is the Musée Guimet, recently removed from Lyons to Paris, which is intended to illustrate the history of religious ceremonial among all races of men.

Other good examples of this class are some of those in Paris, such as the Musée de Marine, which shows not only the development of the merchant and naval marines of the country, but also, by trophies and other historical souvenirs, the history of the naval battles of the nation.

The Musée d'Artillerie does for war, but less thoroughly, what the Marine Museum does in its own department, and there are similar museums in other countries.

Historical museums are manifold in character, and of necessity local in interest. Some relate to the history of provinces or cities. One of the oldest and best of these is the Märkisch Provinzial Museum in Berlin. Many historical societies have collections of this character.

There are museums which illustrate the history of particular towns, events, and individuals. The museum of the city of Paris, in the Hôtel des

Invalides, is one of these. The museum of the Hohenzollerns, in Berlin, contains interesting mementos of the reigning family of Germany. The cathedrals of southern Europe, and St. Paul's, in London, are in some degrees national or civic museums. The Galileo Museum in Florence, the Shakespeare Museum at Stratford, are good examples of the museums devoted to the memory of representative men and the Monastery of St. Mark, in Florence, does as much as could be expected of any museum for the life of Savonarola. The Soane Museum in London, the Thorvaldsen Museum in Copenhagen, are similar in purpose and result, but they are rather biographical than historical. There are also others which illustrate the history of a race, as the Bavarian National Museum in Nuremberg.

The study of civilization or the history of culture and of the developments of the various arts and industries have brought into being special collections which are exceedingly significant and useful. Doctor Klemm and General Pitt-Rivers, in England, were pioneers in the founding of collections of this kind, and their work is permanently preserved in the Museum für Völkerkunde, in Leipzig and at the University of Oxford.

Nearly every museum which admits ethnological material is doing something in this direction. There are a number of beginnings of this sort in this very building.

The best of the art museums are historically arranged and show admirably the development of the pictorial and plastic arts—some, like that in Venice, for a particular school; some that of a country, some that of different countries side by side.

The art museum, it need scarcely be said, contains, more than any other, the materials which I should like to see utilized in the historical museum.

Incidentally or by direct intention, a large collection of local paintings, such as those in Venice or Florence, brings vividly into mind the occurrences of many periods of history, not only historical topography—the architecture, the utensils, weapons, and other appurtenances of domestic, military, ecclesiastical, and governmental routine—but the men and women who made the history, the lowest as well as the most powerful, and the very performers of the deeds themselves, the faces bearing the impress of the passions by which they were moved.

These things are intelligible to those who are trained to observe them. To others they convey but half the lesson they might, or mayhap only a very small part indeed.

The historical museums now in existence contain, as a rule, chance accumulations, like too many natural-history museums of the present, like all in the past. I do not mean any disrespect by the word *chance*, but simply that, though the managers are willing to expend large sums for any specimens which please them, many most instructive ones have been excluded by some artificial limitation. The National Portrait Gallery in London is an instance. Many illustrious men are not represented upon its walls solely because no contemporary pictures of theirs, reaching a certain ideal standard of merit, are in existence.

So, also, the collection of musical instruments at South Kensington, which admits no specimen which is devoid of artistic suggestions—thus barring out the rude and primitive forms which would give added interest to all. The naturalist's axiom, "any specimen is better than no specimen," should be borne in mind in the formation of historical museums, if not rigidly enforced.

Another source of weakness in all museums is one to which attention has already been directed, namely, that they have resigned, without a struggle, to the library material invaluable for the completion of their exhibition series. Pictures are quite as available for museum work as specimens, and it is unwise to leave so many finely illustrated books, lost to sight and memory, on the shelves of the libraries.

That libraries can do good work through the adoption of museum methods has been clearly shown in the British Museum in the exceedingly instructive collections which have of late years been exhibited by its librarians, to illustrate such subjects as the lives of Luther and Michael Angelo, and by their permanent display of pictures and documents referring to the history of London.

The Dyce-Forster collection of autograph documents, letters, and manuscripts is also, in its own way, suggestive. Every large library has done something of this kind in its own way. It remains for some student of history to work out upon a generous plan, and with plenty of exhibition space at his command, the resources which are already in the possession of some great treasure-house like the British Museum.

What the limitations of historical museums are to be it is impossible at present to predict. In museum administration experience is the only safe guide. In the scientific museum many things have been tried, and many things are known to be possible. In the historical museum most of this experimental administration still remains to be performed. The principal object of this communication is to call attention to the general direction in which experiment should be made.

The only safe course to be pursued in the development of plans in any untried department of museum work is to follow the advice which the Apostle Paul proffered to the Thessalonians:

"Prove all things; hold fast that which is good!"

THE MUSEUMS OF THE FUTURE[1]

By George Brown Goode

Assistant Secretary, Smithsonian Institution, in charge of the U. S. National Museum

There is an Oriental saying that the distance between ear and eye is small, but the difference between hearing and seeing very great.

More terse and not less forcible is our own proverb, "To see is to know," which expresses a growing tendency in the human mind.

In this busy, critical, and skeptical age each man is seeking to know all things, and life is too short for many words. The eye is used more and more, the ear less and less, and in the use of the eye, descriptive writing is set aside for pictures, and pictures in their turn are replaced by actual objects. In the schoolroom the diagram, the blackboard, and the object lesson, unknown thirty years ago, are universally employed. The public lecturer uses the stereopticon to reenforce his words, the editor illustrates his journals and magazines with engravings a hundred-fold more numerous and elaborate than his predecessor thought needful, and the merchant and manufacturer recommend their wares by means of vivid pictographs. The local fair of old has grown into the great exposition, often international and always under some governmental patronage, and thousands of such have taken place within forty years, from Japan to Tasmania, and from Norway to Brazil.

Amid such tendencies, the museum, it would seem, should find congenial place, for it is the most powerful and useful auxiliary of all systems of teaching by means of object lessons.

The work of organizing museums has not kept pace with the times. The United States is far behind the spirit of its own people, and less progressive than England, Germany, France, Italy, or Japan. We have, it is true, two or three centers of great activity in museum work, but there have been few new ones established within twenty years, and many of the old ones are in a state of torpor. This can not long continue. The museum of the past must be set aside, reconstructed, transformed from a cemetery of bric-a-brac into a nursery of living thoughts. The museum of the future must stand side by side with the library and the laboratory, as a part of the teaching equipment of the college and university, and in the great cities cooperate with the public library as one of the principal agencies for the enlightenment of the people.

The true significance of the word museum may best be appreciated through an allusion to the ages which preceded its origin—when our ancestors, hundreds of generations removed, were in the midst of those great migrations which peopled Europe with races originally seated in central Asia.

It has been well said that the early history of Greece is the first chapter in the political and intellectual life of Europe. To the history of Greece let us go for the origin of the museum idea, which, in its present form, seems to have found its only congenial home among the European offshoots of the great Indo-Germanic or Aryan division of the world's inhabitants. Long centuries before the invention of written languages there lived along the borders of northern Greece, upon the slopes of Mount Olympus and Helicon, a people whom the later Greeks called "Thracians," a half-mythical race, whose language even has perished. They survived in memory, we are told, as a race of bards, associated with that peculiar legendary poetry of pre-Homeric date, in which the powers of nature were first definitely personified. This poetry belonged, presumably, to an age when the ancestors of the Greeks had left their Indo-European home, but had not yet taken full possession of the lands which were afterwards Hellenic. The spirits of nature sang to their sensitive souls with the voice of brook and tree and bird, and each agency or form which their senses perceived was personified in connection with a system of worship. There were spirits in every forest or mountain, but in Thrace alone dwelt the Muses—the spirits who know and who remember, who are the guardians of all wisdom, and who impart to their disciples the knowledge and the skill to write.

Museums, in the language of Ancient Greece, were the homes of the Muses. The first were in the groves of Parnassus and Helicon, and later they were temples in various parts of Helles. Soon, however, the meaning of the word changed, and it was used to describe a place of study, or a school. Athenæus, in the second century, described Athens as "the museum of Greece," and the name was applied to that portion of the palace of Alexandria which was set apart for the study of the sciences and which contained the famous Alexandrian library. The museum of Alexandria was a great university, the abiding place of men of science and letters, who were divided into many companies or colleges, for the support of each of which a handsome revenue was allotted.

The Alexandrian museum was burned in the days of Cæsar and Aurelian, and the term museum, as applied to a great public institution, dropped out of use from the fourth to the seventeenth century. The disappearance of a word is an indication that the idea for which it stood had also fallen into disfavor, and such, indeed, was the fact.

The history of museum and library runs in parallel lines. It is not until the development of the arts and sciences has taken place, until an extensive written literature has grown up, and a distinct literary and scientific class has been developed, that it is possible for the modern library and museum to come into existence. The museum of the present is more unlike its old-time representative than is our library unlike its prototype.

There were, in the remote past, galleries of pictures and sculpture as well as museums, so called. Public collections of paintings and statuary were founded in Greece and Rome at a very early day. There was a gallery of paintings (Pinacotheca) in one of the marble halls of the Propylæum at Athens, and in Rome there was a lavish public display of works of art.

M. Dezobry, in his brilliant work upon "Rome in the time of Augustus" (*Rome au siècle d'Auguste*), described this phase of the Latin civilization in the first century before Christ.

"For many years," remarks one of his characters, "the taste for paintings has been extending in a most extraordinary manner. In former times they were only to be found in the temples, where they were placed, less for purposes of ornament than as an act of homage to the gods; now they are everywhere, not only in temples, in private houses, and in public halls, but also on outside walls, exposed freely to air and sunlight. Rome is one great picture gallery; the Forum of Augustus is gorgeous with paintings, and they may be seen also in the Forum of Cæsar, in the Roman Forum, under the peristyles of many of the temples, and especially in the porticos used for public promenades, some of which are literally filled with them. Thus everybody is enabled to enjoy them, and to enjoy them at all hours of the day."

The public men of Rome at a later period in its history were no less mindful of the claims of art. They believed that the metropolis of a great nation should be adorned with all the best products of civilization. We are told by Pliny that when Cæsar was dictator, he purchased for 300,000 deniers two Greek paintings, which he caused to be publicly displayed, and that Agrippa placed many costly works of art in a hall which he built and bequeathed to the Roman people. Constantine gathered together in Constantinople the paintings and sculptures of the great masters, so that the city before its destruction became a great museum like Rome.

The taste for works of art was in the days of the ancient civilizations generally prevalent throughout the whole Mediterranean region, and there is abundant reason to believe that there were prototypes of the modern museum in Persia, Assyria, Babylonia, and Egypt, as well as in Rome.

Collections in natural history also undoubtedly existed, though we have no positive descriptions of them. Natural curiosities, of course, found their way into the private collections of monarchs, and were doubtless also in use for study among the savants in the Alexandrian museums. Aristotle, in the fourth century before Christ, had, it is said, an enormous grant of money for use in his scientific researches, and Alexander the Great, his patron, "took care to send to him a great variety of zoological specimens, collected in the countries which he had subdued," and also "placed at his disposal several thousand persons, who were occupied in hunting, fishing, and making the observations which were necessary for completing his History of Animals." If human nature has not changed more than we suppose, Aristotle must have had a great museum of natural history.

When the Roman capital was removed to Byzantium, the arts and letters of Europe began to decline. The church was unpropitious, and the invasions of the northern barbarians destroyed everything. In 476, with the close of the Western Empire, began a period of intellectual torpidity which was to last for a thousand years. It was in Bagdad and Cordova that science and letters were next to be revived, and Africa was to surpass Europe in the exhibit of its libraries.

With the renaissance came a period of new life for collectors. The churches of southern Europe became art galleries, and monarchs and noblemen and ecclesiastical dignitaries collected books, manuscripts, sculptures, pottery, and gems, forming the beginnings of collections which have since grown into public museums. Some of these collections doubtless had their first beginnings in the midst of the Dark Ages within the walls of feudal castles or the larger monasteries, but their number was small, and they must have consisted chiefly of those objects so nearly akin to literature as especially to command the attention of bookish men.

The idea of a great national museum of science and art was first worked out by Lord Bacon in his New Atlantis, a philosophical romance, published at the close of the seventeenth century.

The first scientific museum actually founded was that begun at Oxford, in 1677, by Elias Ashmole, still known as the Ashmolean Museum, composed chiefly of natural-history specimens, collected by the botanists Tradescant, father and son, in Virginia and in the north of Africa. Soon after, in 1753, the British Museum was established by act of Parliament, inspired by the will of Sir Hans Sloane, who, dying in 1749, left to the nation his invaluable collection of books, manuscripts, and curiosities.

Many of the great national museums of Europe had their origin in the private collections of monarchs. France claims the honor of having been the first to change a royal into a national museum when, in 1789, the Louvre came into the possession of a republican government.

It is very clear, however, that democratic England stands several decades in advance—its act, moreover, being one of deliberate founding rather than a species of conquest. A century before this, when Charles I was beheaded by order of Parliament, his magnificent private collection was dispersed. What a blessing it would be to England to-day if the idea of founding a national museum had been suggested to the Cromwellians. The intellectual life of America is so closely bound to that of England, that the revival of interest in museums and in popular education at the middle of the present century is especially significant to us.

The great exhibition of 1851 was one of the most striking features of the industrial revolution in England, that great transformation which, following closely upon the introduction of railroads, turned England, feudal and agricultural, into England democratic and commercial. This exhibition marked an epoch in the intellectual progress of English-speaking peoples. "The great exhibition," writes a popular novelist—a social philosopher as well—"did one great service for country people: It taught them how easy it is to get to London, and what a mine of wealth, especially for after-memory and purposes of conversation, exists in that great place."

Our own Centennial Exhibition in 1876 was almost as great a revelation to the people of the United States. The thoughts of the country were opened to many things before undreamed of. One thing we may regret, that we have no such widespread system of museums as that which has developed in the motherland, with South Kensington as its administrative center.

Under the wise administration of the South Kensington staff, an outgrowth of the events of 1851, a great system of educational museums has been developed all through the United Kingdom. A similar extension of public museums in this country would be quite in harmony with the spirit of the times, as shown in the present efforts toward university extensions.

England has had nearly forty years in which to develop these tendencies and we but thirteen since our exhibition. May we not hope that within a like period of time and before the year 1914 the United States may have attained the position which England now occupies, at least in the respect of popular interest and substantial governmental support.

There are now over one hundred and fifty public museums in the United Kingdom, all active and useful. The museum systems of Great Britain are, it seems to me, much closer to the ideal which America should follow than are those of either France or Germany. They are designed more thoughtfully to meet the needs of the people, and are more intimately intertwined with the policy of national, popular education. Sir Henry Cole, the founder of the "department of science and art," speaking of the purpose of the museum under his care, said to the people of Birmingham in 1874: "If you wish your schools of science and art to be effective, your health, the air, and your food to be wholesome, your life to be long, your manufactures to improve, your trade to increase, and your people to be civilized, you must have museums of science and art, to illustrate the principles of life, health, nature, science, art, and beauty."

Again, in words as applicable to America of to-day as to Britain in 1874, said he: "A thorough education and a knowledge of science and art are vital to the nation and to the place it holds at present in the civilized world. Science and art are the lifeblood of successful production. All civilized nations are running a race with us, and our national decline will date from the period when we go to sleep over the work of education, science, and art. What has been done is at the mere threshold of the work yet to be done."

The museums of the future in this democratic land should be adapted to the needs of the mechanic, the factory operator, the day laborer, the salesman, and the clerk, as much as to those of the professional man and the man of leisure. It is proper that there be laboratories and professional libraries for the development of the experts who are to organize, arrange, and explain the museums. It is proper that the laboratories be utilized to the fullest extent for the credit of the institution to which they belong. No museum can grow and be respected which does not each year give additional proofs of its claims to be considered a center of learning.

On the other hand, the public have a right to ask that much shall be done directly in their interest. They will gladly allow the museum officer to use part of his time in study and experiment. They will take pride in the possession by the museum of tens of thousands of specimens, interesting only to the specialists, hidden away perpetually from public view, but necessary for purpose of scientific research. These are foundations of the intellectual superstructure which gives the institution its standing.

Still, no pains must be spared in the presentation of the material in the exhibition halls. The specimens must be prepared in the most careful and artistic manner, and arranged attractively in well-designed cases and behind the clearest of glass. Each object must bear a label, giving its name and history so fully that all the probable questions of the visitor are answered in advance. Books of reference must be kept in convenient places. Colors of walls, cases, and labels must be restful and quiet, and comfortable seats should be everywhere accessible, for the task of the museum visitor is a weary one at best.

In short, the public museum is, first of all, for the benefit of the public. When the officers are few in number, each must of necessity devote a considerable portion for his time to the public halls. When the staff becomes larger, it is possible by specialization of work to arrange that certain men may devote their time uninterruptedly to laboratory work, while others are engaged in the increase of the collections and their installation.

I hope and firmly believe that every American community with inhabitants to the number of five thousand or more will within the next half century have a public library, under the management of a trained librarian. Be it ever so small, its influence upon the people would be of untold value. One of the saddest things in this life is to realize that in the death of the elder members of a community so much that is precious in the way of knowledge and experience is lost to the world. It is through the agency of books that mankind benefits by the toil of past generations and is able to avoid their errors.

In these days, when printing is cheap and authors are countless, that which is good and true in human thought is in danger of being entirely overlooked. The daily papers, and above all the overgrown and uncanny Sunday papers, are like weeds in a garden, whose rank leaves not only consume the resources of the soil but hide from view the more modest and more useful plants of slower growth.

Most suggestive may we find an essay on Capital and Culture in America, which recently appeared in one of the English reviews. The author, a well-known Anglo-American astronomer, boldly asserts that—

Year by year it becomes clearer that, despite the large absolute increase in the number of men and women of culture in America, the nation is deteriorating in regard to culture. Among five hundred towns where formerly courses of varied entertainments worthy of civilized communities—concerts, readings, lectures on artistic, literary, and scientific subjects, etc.—were successfully arranged season after season, scarcely fifty now feel justified in continuing their efforts in the cause of culture, knowing that the community will no longer support them. Scientific, literary, and artistic societies, formerly flourishing, are now dying, or dead in many cities which have in the meantime increased in wealth and population.

He instances Chicago as typical of an important portion of America, and cites evidences of decided deterioration within sixteen years.

The people's museum should be much more than a house full of specimens in glass cases. It should be a house full of ideas, arranged with the strictest attention to system.

I once tried to express this thought by saying, "An efficient educational museum may be described as a collection of instructive labels, each illustrated by a well-selected specimen."

The museum, let me add, should be more than a collection of specimens well arranged and well labeled. Like the library, it should be under the constant supervision of one or more men well informed, scholarly, and withal practical, and fitted by tastes and training to aid in the educational work.

I should not organize the museum primarily for the use of the people in their larval or school-going stage of existence. The public-school teacher, with the illustrated text-book, diagrams, and other appliances, is in these days a professional outfit which is usually quite sufficient to enable him to teach his pupils. School days last, at the most, only from five to fifteen years, and they end with the majority of mankind before their minds have reached the stage of growth most favorable for the reception and assimilation of the best and most useful thought. Why should we be crammed in the times of infancy and kept in a state of mental starvation during the period which follows, from maturity to old age, a state which is disheartening and unnatural, all the more because of the intellectual tastes which have been stimulated and partially formed by school life.

The boundary line between the library and the museum is neither straight nor plain. The former, if its scope be rightly indicated by its name, is primarily a place for books. The latter is a depository for objects of every kind, books not excepted.

The British Museum, with its libraries, its pictures, its archæological galleries, its anthropological, geological, botanical, and zoological collections, is an example of the most comprehensive interpretation of the term.

Professor Huxley has described the museum as "a consultative library of objects." This definition is suggestive but unsatisfactory. It relates only to the contents of the museum, as distinguished from those of the library, and makes no reference to the differences in the methods of their administration. The treasures of the library must be examined one at a time and by one person at a time; their use requires long-continued attention and their removal from their proper places in the system of arrangement. Those of the museum are displayed to public view, in groups, in systematic sequence, so that they have a collective as well as an individual significance. Furthermore, much of their meaning may be read at a glance.

The museum cultivates the powers of observation, and the casual visitor even makes discoveries for himself and under the guidance of the labels forms his own impressions. In the library one studies the impressions of others. The library is most useful to the educated, the museum to educated and uneducated alike, to the masses as well as to the few, and is a powerful stimulant to intellectual activity in either class. The influence of the museum upon a community is not so deep as that of the library, but extends to a much larger number of people.

The National Museum has 300,000 visitors a year, each of whom carries away a certain number of new thoughts.

The two ideas may be carried out, side by side, in the same building,

and if need be under the same management, not only without antagonism, but with advantage.

That the proximity of a good library is absolutely essential to the
usefulness of a museum will be admitted by everyone.

I am confident also that a museum, wisely organized and properly
arranged, is certain to benefit the library near which it stands in many
ways through its power to stimulate interest in books, thus increasing the
general popularity of the library and enlarging its endowment.

Many books, and valuable ones, would be required in the first kind of
museum work, but it is not intended to enter into competition with the
library. (When necessary, volumes could be duplicated.) It is very often
the case, however, that books are more useful and safer in the museum
than on the library shelves, for in the museum they may be seen daily by
thousands, while in the library their very existence is forgotten by all
except their custodian.

Audubon's Birds of North America is a book which everyone has heard
of and which everyone wants to see at least once in his lifetime. In a
library, it probably is not examined by ten persons in a year; in a museum,
the volumes were exposed to view in a glass case, a few of the most
striking plates attractively framed and hung upon the wall near at hand,
it teaches a lesson to every passer-by.

The library may be called upon for aid by the museum in many directions. Pictures are often better than specimens to illustrate certain ideas.
The races of man and their distribution can only be shown by pictures
and maps. Atlases of ethnological portraits and maps are out of place in
a library if there is a museum near by in which they can be displayed.
They are not even members of the class described by Lamb as "books
which are not books." They are not books, but museum specimens, mas-
querading in the dress of books.

There is another kind of depository which, though in external features
so similar to the museum, and often confused with it in name as well as

in thought, is really very unlike it. This is the art gallery. The scientific tendencies of modern thought have permeated every department of human activity, even influencing the artist. Many art galleries are now called museums, and the assumption of the name usually tends toward the adoption in some degree of a scientific method of installation. The difference between a museum and a gallery is solely one of method of management. The Musée des Thermes—the Cluny Museum—in Paris is, notwithstanding its name, simply a gallery of curious objects. Its contents are arranged primarily with reference to their effect. The old monastery in which they are placed affords a magnificent example of the interior decorative art of the Middle Ages.

The Cluny Museum is a most fascinating and instructive place. I would not have it otherwise than it is, but it will always be unique, the sole representative of its kind. The features which render it attractive would be ruinous to any museum. It is, more than any other that I know, a collection arranged from the standpoint of the artist. The same material, in the hands of a Klemm or a Pitt-Rivers, arranged to show the history of human thought, would, however, be much more interesting, and, if the work were judiciously done, would lose none of its æsthetic allurements.

Another collection of the same general character as the one just described is the Soane Museum in London. Another, the famous collection of crown jewels and metal work in the Green Vaults at Dresden, a counterpart of which may be cited in the collection in the Tower of London. The Museum of the Hohenzollerns in Berlin and the Museum of the City of Paris are of necessity unique. Such collections can not be created. They grow in obedience to the action of natural law, just as a tree or a sponge may grow.

The city which is in the possession of such an heirloom is blessed just as is the possessor of a historic surname or he who inherits the cumulative genius of generations of gifted forefathers. The possession of one or a score of such shrines does not, however, free any community from the obligation to form a museum for purposes of education and scientific research.

The founding of a public museum in a city like Brooklyn is a work whose importance can scarcely be overestimated. The founders of institutions of this character do not often realize how much they are doing for the future. Opportunity such as that which is now open to the members of the Brooklyn Institute occur only once in the lifetime of a nation. It is by no means improbable that the persons now in this room have it in their power to decide whether, in the future intellectual progress of this nation, Brooklyn is to lead or to follow far in the rear.

Many of my hearers are doubtless familiar with that densely populated wilderness, the east end of London, twice as large as Brooklyn, yet with scarce an intellectual oasis in its midst. Who can say how different might have been its condition to-day if Walter Besant's apostolic labors had begun a century sooner, and if the People's Palace, that wonderful materialization of a poet's dream, had been for three generations brightening the lives of the citizens of the Lower Hamlets and Hackney?

Libraries and museums do not necessarily spring up where they are needed. Our governments, Federal, State, and municipal, are not "paternal" in spirit. They are less so even in practical working than in England, where, notwithstanding the theory that all should be left to private effort, the Government, under the leadership of the late Prince Consort and of the Prince of Wales, has done wonderful things for all the provincial cities, as well as for London, in the encouragement of libraries, museums, art, and industrial education.

However much the state may help, the private individual must lead, organize, and prepare the way. "It is universally admitted," said the Marquis of Lansdowne in 1847, "that governments are the worst of cultivators, the worst of manufacturers, the worst of traders," and Sir Robert Peel said in similar strain that "the action of government is torpid at best."

In beginning a museum the endowment is of course the most essential thing, especially in a great city like Brooklyn, which has a high ideal of what is due to the intelligence of its populace and to the civic dignity.

Unremunerated service in museum administration, though it may be enthusiastically offered and conscientiously performed, will in the end fail to be satisfactory. Still more is it impossible for a respectable museum to grow up without liberal expenditure for the acquisition of collections and their installation.

Good administration is not to be had for nothing. As to the qualification of a museum administrator, whether it be for a museum of science or a museum of art, it is perhaps superfluous to say that he should be the very best obtainable, a man of ability, enthusiasm, and, withal, of experience; for the administration of museums and exhibitions has become of late years a profession, and careful study of methods of administration is indispensable. If the new administrator has not had experience he must needs gain it at the expense of the establishment which employs him— an expense of which delay, waste, and needless experiment form considerable elements.

No investment is more profitable to a museum than that in the salary fund. Around a nucleus of men of established reputation and administrative tact will naturally grow up a staff of volunteer assistants whose work, assisted and directed in the best channels, will be of infinite value.

The sinews and brains of the organism being first provided, the development of its body still remains. The outer covering, the dress, can wait. It is much better to hire buildings for temporary use, or to build rude fireproof sheds, than to put up a permanent museum building before at least a provisional idea of its personnel and contents has been acquired.

As has been already said, a museum must spend money in the acquisition of collections, and a great deal of money. The British Museum has already cost the nation for establishment and maintenance not far from $30,000,000. Up to 1882 over $1,500,000 had been expended in purchase of objects for the art collections at South Kensington alone.

Such expenditures are usually good investments of national funds, however. In 1882, after about twenty-five years of experience, the buildings and contents of the South Kensington Museum had cost the nation about $5,000,000, but competent authorities were satisfied that an auction on the premises could not bring less than $100,000,000. For every dollar spent, however, gifts will come in to the value of many dollars. In this connection it may not be amiss to quote the words of one of the most experienced of English museum administrators (presumably Sir Philip Cunliffe Owen) when asked many years ago whether Americans might not develop great public institutions on the plan of those at Kensington:

Let them plant the thing [he said], and it can't help growing, and most likely beyond their powers—as it has been almost beyond ours—to keep up with it. What is wanted first of all is one or two good good brains, with the means of erecting a good building on a piece of ground considerably larger than is required for that building. Where there have been secured substantial, luminous galleries for exhibition, in a fireproof building, and these are known to be carefully guarded by night and day, there can be no need to wait long for treasures to flow into it. Above all, let your men take care of the interior and not set out wasting their strength and money on external grandeur and decoration. The inward built up rightly, the outward will be added in due season.[2]

Much will, of course, be given to any museum which has the confidence of the public—much that is of great value, and much that is useless.

The Trojans of old distrusted the Greeks when they came bearing gifts. The museum administrator must be on his guard against every one who proffers gifts. An unconditional donation may be usually accepted without hesitation, but a gift coupled with conditions is, except in very extraordinary cases, far from a benefaction.

A donor demands that his collection shall be exhibited as a whole, and kept separate from all others. When his collection is monographic in character and very complete, it is sometimes desirable to accept it on such conditions. As a rule, however, it is best to try to induce the donor to allow his collections to be merged in the general series—each object being separately and distinctively labeled. I would not be understood to say that the gift of collections is not, under careful management, a most beneficial source of increase to a public collection. I simply wish to call attention to the fact that a museum which accepts without reserve gifts of every description, and fails to reenforce these gifts by extensive and judicious purchasing, is certain to develop in an unsystematical and ill-balanced way.

Furthermore, unless a museum be supported by liberal and constantly increasing grants from some State or municipal treasury, it will ultimately become suffocated. It is essential that every museum, whether of science or art, should from the start make provision for laboratories and storage galleries as well as for exhibition halls.

All intellectual work may be divided into two classes, the one tending toward the increase of knowledge, the other toward its diffusion—the one toward investigation and discovery, the other toward the education of the people and the application of known facts to promoting their material welfare. The efforts of learned men are sometimes applied solely to one of these departments of effort, sometimes to both, and it is generally admitted by the most advanced teachers that, for their students as well as for themselves, the happiest results are reached by investigation and instruction simultaneously. Still more is this true of institutions of learning. The college which imparts only secondhand knowledge to its students belongs to a stage of civilization which is fast being left behind. The museum likewise must, in order to perform its proper functions, contribute to the advancement of learning through the increase as well as through the diffusion of knowledge.

We speak of educational museums and of the educational method of installation so frequently that there may be danger of inconsistency in the use of the term. An educational museum, as it is usually spoken of, is one in which an attempt is made to teach the unprofessional visitor—an institution for popular education by means of labeled collections, and it may be also by popular lectures. A college museum, although used as an aid to advanced instruction, is not an "educational museum" in the ordinary sense; nor does a museum of research, like the Museum of Comparative Zoology in Cambridge, Massachusetts, belong to this class, although to a limited extent it attempts and performs popular educational work in addition to its other functions.

In the National Museum in Washington the collections are divided into two great classes—the exhibition series, which constitutes the educational portion of the Museum, and is exposed to public view with all possible accessions for public entertainment and instruction, and the study series, which is kept in the scientific laboratories, and is scarcely examined except by professional investigators.

In every properly conducted museum the collections must from the very beginning divide themselves into these two classes, and in planning for its administration provision should be made not only for the exhibition of objects in glass cases, but for the preservation of large collections not available for exhibition, to be used for the studies of a very limited number of specialists.

Lord Bacon, who, as we have noticed, was the first to whom occurred the idea of a great museum of science and art, complained three centuries ago, in his book On the Advancement of Learning, that up to that time the means for intellectual progress had been used exclusively for "amusement" and "teaching," and not for the "augmentation of science."

It will undoubtedly be found desirable for certain museums, founded for local effect, to specialize mainly in the direction of popular education. If they can not also provide for a certain amount of scholarly endeavor in connection with the other advantages, it would be of the utmost importance that they should be assorted by a system of administrative cooperation with some institution which is in the position of being a center of original work.

The general character of museums should be clearly determined at its very inception. Specialization and division of labor are essential for institutions as well as for individuals. It is only a great national museum which can hope to include all departments and which can with safety encourage growth in every direction.

A city museum, even in a great metropolis like Brooklyn, should, if possible, select certain special lines of activity and pursue them with the intention of excelling. If there are already beginnings in many directions, it is equally necessary to decide which lines of development are to be favored in preference to all others. Many museums fail to make this choice at the start, and instead of steering toward some definite point, drift hither and thither, and it may be, are foundered in mid-ocean.

There is no reason why the museum of the Brooklyn Institute may not in time attain to world-wide fame and attract students and visitors from afar. It would be wise, perhaps, in shaping its policy to remember that in the twin city of New York are two admirable museums which may be met more advantageously in cooperation than in rivalry. Brooklyn may appropriately have its own museum of art and its museum of natural history, but they should avoid the repetition of collections already so near at hand.

In selecting courses for the development of a museum, it may be useful to consider what are the fields open to museum work.

As a matter of convenience museums are commonly classed in two groups—those of science and those of art, and in Great Britain the great national system is mainly under the control of The Science and Art Department of the Committee of Council on Education.

The classification is not entirely satisfactory, since it is based upon methods of arrangement, rather than upon the nature of the objects to

be arranged, and since it leaves a middle territory (only partially occupied
by the English museum men of either department), a great mass of museum material of the greatest moment both in regard to its interest and
its adaptability for purposes of public instruction.

On the one side stand the natural history collections, undoubtedly
best to be administered by the geologist, botanist, and zoologist. On the
other side are the fine art collections, best to be arranged from an æsthetic standpoint, by artists. Between is a territory which no English
word can adequately describe—which the Germans call *Culturgeschichte*—the natural history of cult, or civilization, of man and his ideas
and achievements. The museums of science and art have not yet learned
how to partition this territory. An exact classification of museums is not
at present practicable, nor will it be until there has been some redistribution of the collections which they contain. It may be instructive, however, to pass in review the principal museums of the world, indicating
briefly their chief characteristics.

Every great nation has its museum of nature. The natural history department of the British Museum, recently removed from the heart of
London to palatial quarters in South Kensington, is probably the most
extensive—with its three great divisions, zoological, botanical, and geological. The Musée d'Histoire Naturelle, in the garden of plants in Paris,
founded in 1795, with its galleries of anatomy, anthropology, zoology,
botany, mineralogy, and geology, is one of the most extensive, but far less
potent in science now than in the days of Cuvier, Lamarck, St. Hilaire,
Jussieu, and Brongniart. In Washington, again, there is a National Museum with anthropological, zoological, botanical, mineralogical, and geological collections in one organization, together with a large additional
department of arts and industries, or technology.

Passing to specialized natural history collections, perhaps the most
noteworthy are those devoted to zoology, and chief among them that in
our own American Cambridge. The Museum of Comparative Zoology,

founded by the Agassizes "to illustrate the history of creation, as far as the present state of knowledge reveals that history," was in 1887 pronounced by the English naturalist, Alfred Russell Wallace, "to be far in advance of similar institutions in Europe as an educational institution, whether as regards the general public, the private student, or the specialist."

Next to Cambridge, after the zoological section of the museums of London and Paris, stands the collections in the Imperial Cabinet in Vienna, and those of the zoological museums in Berlin, Leyden, Copenhagen, and Christiania.

Among botanical museums, that in the Royal Gardens at Kew, near London, is preeminent, with its colossal herbarium, containing the finest collection in the world, and its special museum of economic botany, founded in 1847, both standing in the midst of a collection of living plants. There is also in Berlin the Royal Botanical Museum, founded in 1818 as the Royal Herbarium; in St. Petersburg, the Herbaria of the Imperial Botanical Garden.

Among the geological and mineralogical collections the mineral cabinet in Vienna, arranged in the imperial castle, is among the first.

The Museum of Practical Geology in London, which is attached to the Geological Survey of the United Kingdom, was founded in 1837, to exhibit the collections of the survey, in order to "show the applications of geology to the useful purposes of life." Like every other healthy museum, it soon had investigations in progress in connection with its educational work, and many very important discoveries have been made in its laboratories. It stands in the very first rank of museums for popular instruction, the arrangement of the exhibition halls being most admirable. Of museums of anatomy there are thirty of considerable magnitude, all of which have grown up in connection with schools of medicine and surgery, except the magnificent Army Medical Museum in Washington.

The Medical Museum of the Royal College of Surgeons in London is probably first in importance. The collections of St. Thomas's, Guy's, St. George's, and other hospitals are very rich in anatomical and pathological specimens. The oldest public anatomical museum in London is that of St. Bartholomew's.

Paris, Edinburgh, and Dublin have large anatomical and materia medica collections. As a rule, the medical museums of Europe are connected with universities. Doctor Billings, curator of the Army Medical Museum in Washington, has traced accurately the growth of medical collections both at home and abroad, and from his address upon medical museums, as president of the Congress of American Physicians and Surgeons, delivered in 1888, the facts here stated relating to this class of museums have been gathered. The Army Medical Museum apparently owes its establishment to Doctor William A. Hammond, in 1862. The museum contained in 1888 more than 15,000 specimens, besides those contained in the microscopical department. "An ideal medical museum," says Doctor Billings, "should be very complete in the department of preventive medicine or hygiene. It is a wide field, covering, as it does, air, water, food, clothing, habitations, geology, meteorology, occupations, etc., in their relations to the production or prevention of disease, and thus far has had little place in medical museums, being taken up as a specialty in the half dozen museums of hygiene which now exist."

William Hunter formed the great Glasgow collection between the years 1770 and 1800, and John Hunter, in 1787, opened the famous Hunterian Museum in London, bought by the English Government soon after (1799), and now known as the Museum of the Royal College of Surgeons.

Paris is proud of the two collections at the School of Medicine, the Musée Orfila and the Musée Dupuytren, devoted, the one to normal, the other to pathological anatomy.

Ethnographic museums are especially numerous and fine in the northern part of continental Europe. They were proposed more than half a century ago by the French geographer Jomard, and the idea was first carried into effect about 1840 in the establishment of the Danish Ethnographical Museum, which long remained the best in Europe. Within the past twenty years there has been an extraordinary activity in this direction.

In Germany, besides the museums in Berlin, Dresden, and Leipsic, considerable collections have been founded in Hamburg and Munich. Austria has in Vienna two for ethnography, the Court Museum (Hof-Museum) and the Oriental (Orientalisches) Museum. Holland has reorganized the National Ethnographical Museum (Rijks Ethnographisch Museum) in Leyden, and there are smaller collections in Amsterdam, Rotterdam, and The Hague. France has founded the Trocadero (Musée de Trocadéro). In Italy there is the important Prehistoric and Ethnographic Museum (Museo prehistorico ed ethnografico) in Rome, as well as the collection of the Propaganda, and there are museums in Florence and Venice.

Ethnographical museums have also been founded in Christiania and Stockholm, the latter of which will include the rich material collection by Doctor Stolpe on the voyage of the frigate *Vanadis* around the world. In England there is less attention to the subject—the Christy collection in the British Museum being the only one specially devoted to ethnography, unless we include also the local Blackmore Museum in Salisbury.

In the United States the principal establishments arranged on the ethnographic plan are the Peabody Museum of Archæology in Cambridge and the collections in the Peabody Academy of Sciences in Salem, and the American Museum of Natural History in New York.

The ethnological collections in Washington are classified on a double system; in one of its features corresponding to that of the European; in the other, like the famous Pitt-Rivers collection at Oxford, arranged to show the evolution of culture and civilization without regard to race. This broader plan admits much material excluded by the advocates of ethnographic museums, who devote their attention almost exclusively to the primitive or non-European peoples.

In close relation to the ethnographic museums are those which are devoted to some special field of human thought and interest. Most remarkable among these, perhaps, is the Musée Guimet, recently removed from Lyons to Paris, which is intended to illustrate the history of religious ceremonial among all races of men. Other good examples of this class are some of those in Paris, such as the Musée de Marine, which shows not only the development of the merchant and naval marines of the country, but also, by trophies and other historical souvenirs, the history of the naval battles of the nation. The Musée d'Artillerie does for war, but less thoroughly, what the Marine Museum does in its own department, and there are similar museums in other countries. Of musical museums, perhaps, the most important is the Musée Instrumental founded by Clapisson, attached to the Conservatory of Music in Paris. There is a magnificent collection of musical instruments at South Kensington, but its contents are selected in reference to their suggestiveness in decorative art. There are also large collections in the National Museum in Washington and in the Conservatory of Music in Boston, and the Metropolitan Museum in New York has recently been given a very full collection by Mrs. John Crosby Brown, of that city.

There is a Theatrical Museum at the Académie Française in Paris, a Museum of Journalism in Antwerp, a Museum of Pedagogy in Paris, which has its counterpart in South Kensington. These are professional, rather than scientific or educational, as are perhaps also the Museum of Practical Fish Culture at South Kensington and the Museums of Hygiene in London and Washington.

Archæological collections are of two classes, those of prehistoric and historic archæology. The former are usually absorbed by the ethnographic museums, the latter by the art museums. The value to the historian of archæological collections, both historic and prehistoric, has long been understood. The museums of London, Paris, Berlin, and Rome need no comment. In Cambridge, New York, and Washington are immense collections of the remains of man in America in the pre-Columbian period, collections which are yearly growing in significance, as they are made the subject of investigation, and there is an immense amount of material of this kind in the hands of institutions and private collectors in all parts of the United States.

The museum at Naples shows, so far as a museum can, the history of Pompeii at one period. The museum of St. Germain, near Paris, exhibits the history of France in the time of the Gauls and of the Roman occupation. In Switzerland, especially at Neuchatel, the history of the inhabitants of the Lake Dwellings is shown. The Assyrian and Egyptian galleries in the British Museums are museums of themselves.

Historical museums are manifold in character, and of necessity local in interest. Some relate to the history of provinces or cities. One of the oldest and best of these is the Märkisch Provinzial Museum in Berlin; another is the museum of the city of Paris, recently opened in the Hotel Canaveral. Many historical societies have collections of this character. Some historical museums relate to a dynasty, as the Museum of the Hohenzollerns in Berlin.

The cathedrals of southern Europe, and St. Paul's, in London, are in some degrees national or civic museums. The Galileo Museum in Florence, the Shakespeare Museum at Stratford, are good examples of the museums devoted to the memory of representative men, and the Monastery of St. Mark, in Florence, does as much as could be expected of any museum for the life of Savonarola. The Soane Museum in London, the Thorvaldsen Museum in Copenhagen, are similar in purpose and result, but they are rather biographical than historical. There are also others which illustrate the history of a race, as the Bavarian National Museum in Nuremberg.

The museums of fine art are the most costly and precious of all, since they contain the masterpieces of the world's greatest painters and sculptors. In Rome, Florence, Venice, Naples, Bologna, Parma, Milan, Turin, Modena, Padua, Ferrara, Brescia, Sienna, and Pisa; in Munich, Berlin, Dresden, Vienna, and Prague; in Paris, and many provincial cities of France; in London, St. Petersburg, Madrid, Copenhagen, Brussels, Antwerp, and The Hague, are great collections, whose names are familiar to us all, each the depository of priceless treasures of art. Many of these are remarkable only for their pictures and statuary, and might with equal right be called picture galleries; others abound in the minor products of artists, and are museums in the broader sense.

Chief among them is the Louvre, in Paris, with its treasures worth a voyage many times around the world to see; the Vatican, in Rome, with its three halls of antique sculptures, its Etruscans, Egyptian, Pagan, and Christian museums, its Byzantine gallery and its collection of medals; the Naples Museum (Musée di Studii) with its marvelous Pompeiian series; the Uffizi Museum in Florence, overflowing with paintings and sculptures, ancient and modern, drawings, engraved gems, enamels, ivories, tapestries, medals, and works of decorative art of every description.

There are special collections on the boundary line between art and ethnology, the manner of best installation for which has scarcely yet been determined. The Louve admits within its walls a museum of ship models (Musée de Marine). South Kensington includes musical instruments, and many other objects equally appropriate in an ethnological collection. Other art museums take up arms and armor, selected costumes, shoes, and articles of household use. Such objects, like porcelains, laces, medals, and metal work, appeal to the art museum administrator through their decorations and graceful forms. For their uses he cares presumably nothing. As a consequence of this feeling only articles of artistic excellence have been saved, and much has gone to destruction which would be of the utmost importance to those who are now studying the history of human thought in the past.

On the other hand, there is much in art museums which might to much better purpose be delivered to the ethnologist for use in his exhibition cases. There is also much which the art museums, tied as they often are to traditionary methods of installation, might learn from the scientific museums.

Many of the arrangements in the European art collections are calculated to send cold shivers down the back of a sensitive visitor. The defects of these arrangements have been well described by a German critic, W. Bürger.

Our museums [he writes] are the veritable graveyards of arts, in which have been heaped up, with a tumulous-like promiscuousness, the remains which have been carried thither. A Venus is placed side by side with a Madonna, a satyr next to a saint. Luther is in close proximity to a Pope, a painting of a lady's chamber next to that of a church. Pieces executed for churches, palaces, city halls, for a particular edifice, to teach some moral or historical truth, designed for some especial light, for some well-studied surrounding, all are hung pellmell upon the walls of some noncommittal gallery—a kind of posthumous asylum, where a people, no longer capable of producing works of art, come to admire this magnificent gallery of débris.

When a museum building has been provided, and the nucleus of a collection and an administrative staff are at hand, the work of museum-building begins, and this work, it is to be hoped, will not soon reach an end. A finished museum is a dead museum, and a dead museum is a useless museum. One thing should be kept prominently in mind by any organization which intends to found and maintain a museum, that the work will never be finished; that when the collections cease to grow, they begin to decay. A friend relating an experience in South Kensington, said: "I applied to a man who sells photographs of such edifices for pictures of the main building. He had none. 'What, no photographs of the South

Kensington Museum!' I exclaimed, with some impatience. 'Why, sir,' replied the man mildly, 'you see the museum doesn't stand still long enough to be photographed.' And so indeed it seems," continued Mr. Conway, "and this constant erection of new buildings and of new decorations on those already erected, is the physiognomical expression of the new intellectual and æsthetic epoch which called the institution into existence, and is through it gradually climbing to results which no man can foresee."

My prayer for the museums of the United States and for all other similar agencies of enlightenment is this—that they may never cease to increase.

NOTES

GEORGE BROWN GOODE 1851–1896

1. The author gratefully acknowledges permission to use parts of "History in a Natural History Museum: George Brown Goode and the Smithsonian Institution," in *The Public Historian*, 10: 1988, 7–26. Basic biographical information is available in a series of eulogies in *A Memorial of George Brown Goode*, which was printed in the *Annual Report of the Regents of the Smithsonian Institution: Report of the U. S. National Museum*, Part II (Washington, D.C., 1901). More recently his contributions to museum administration have been examined in Edward P. Alexander, *Museum Masters: Their Museums and Their Influence* (Nashville: American Association for State and Local History, 1983), pp. 277–310; and his historical work is noted in an essay by G. Carroll Lindsay in Clifford L. Lord, ed., *Keepers of the Past* (Chapel Hill: University of North Carolina Press, 1965). Langley's "Memoir of George Brown Goode," in the *Memorial of Goode*, cites anecdotal information about Goode's early years and makes references to family correspondence not found by this author.

2. Numerous observations and phrases from the essays reprinted in this volume will not be included in the endnotes.

3. Ralph W. Dexter discusses tensions among some of these students in "The 'Salem Secession' of Agassiz Zoologists," *Essex Institute Historical Collections,* 101 (1965): 27–39. Goode was listed simply as a "scientific student" in Lawrence Scientific School in the *Harvard University Catalogue, 1870–1871,* p. 75.

4. Goode, as curator, contributed information on the development of the museum to the *Annual Reports to the Board of Trustees of Wesleyan University* from 1871 until he resigned in 1879. A brief sketch of Goode appeared in the *Wesleyan Argus,* 30 (October 20, 1896): 17–19. Carl Price, *Wesleyan's First Century* (Middletown: Wesleyan University, 1932), pp. 110–111. Also see S. G. Kohlstedt, "Museums on Campus: A Tradition of Inquiry and Teaching," in Ronald Rainger, Keith Benson, and Jane Maienschein, eds., *The American Development of Biology* (Philadelphia: University of Pennsylvania Press, 1988), pp. 15–47.

5. Goode to Baird, esp. 29 April 1972, 18 September 1873, and 17 October 1873, Baird MSS, RU 52, Smithsonian Institution Archives, Washington, D.C. (hereafter SIA). For a discussion of the collections and donors see the "Visitors Guide to the Museum" reprinted as Appendix B in the *Seventh Annual Report of the Curators of the Museum of Wesleyan University* (Middletown, Conn.: n.p., 1877), pp. 13–24.

6. For the 1876 International Exhibition he chose to develop a new classification system that organized zoological and anthropological information in relationship to human experience and development; see his *Classification of the Collection to Illustration the Animal Resources of the United States* (Washington, D.C., 1876). Correspondence relating to this effort are also found in a small set of Goode papers in the Manuscript Division of the Library of Congress. The Smithsonian exhibit so impressed Frederick Starr that he sketched out the labels in his manuscript journal, p. 57; see Starr MSS, Regenstein Library, University of Chicago. Joseph Henry, Secretary of the Smithsonian, also issued a *Circular* on June 10, 1876, to solicit contributions from exhibitors. The 1876 Exposition was revived in 1976 by a major exhibit in the Arts and Industries Building of the Smithsonian Institution and discussed in a catalogue edited by Robert Post, *The Centennial Exhibition: A Treatise upon Selected Aspects of the Great International Exhibition Held in Philadelphia on the Occasion of our Nation's One Hun-*

dredth Birthday . . . . (Washington, D.C.: National Museum of History and Technology, 1976).

7. Goode's work was subject, however, to Baird's approval, as suggested in the detailed letters which passed between the two men when the building and cases were under construction in 1881; see RU 112, Assistant Secretary, SIA.

8. See the series of letters from Goode to Baird in April and May of 1880; Baird MSS, RU 7002, SIA, Washington, D.C.

9. His "plan of operation and a philosophical system of classification" for museum operation was outlined in the first *Annual Report* of the United States National Museum. Despite a critical review in *Science* [2 (July 20 and August 3, 1883): 63–66 and 119–123], this report established Goode as a foremost American theorist on museum practice.

10. Goode's outlook on anthropology, in particular, is sketched in Robert W. Rydell, *All the World's a Fair: Visions of Empire at American International Expositions, 1876–1916* (Chicago: University of Chicago Press, 1984), esp. pp. 43–46.

11. On intellectual life in Washington, see J. Kirkpatrick Flack, *Desideratum in Washington: The Intellectual Community in the Capital City, 1870–1900* (Cambridge: Schenkman Pub. Co., 1975); and Michael Lacey, "Mysteries of Earth-Building Dissolve: A Study of Washington's Intellectual Community and the Origins of Environmentalism in the Late Nineteenth Century" (Ph.D. dissertation: George Washington University, 1979). Among other things, Goode helped to develop the area of Lanier Heights where he built his own home in 1880; see James Goode, *Capital Losses: A Cultural History of Washington's Destroyed Buildings* (Washington: Smithsonian Institution Press, 1979), pp. 86–87.

12. T. H. Bean to Goode, 12 September 1884, RU 201, Assistant Secretary, SIA.

13. On Goode's personal life, see sources in n. 1 above. Occasionally he provided a glimpse of himself, as in a long letter addressed to Dr. Hamlin, undated, that pointed out his own capacity to reconcile religion and science. See Goode Collection, RU 7050, SIA. William Cox to Spencer F. Baird, 29 July 1886, Baird MSS, RU 7002, SIA.

14. Elizabeth Agassiz and Louis Agassiz, *Seaside Studies in Natural History*, p. 25, and cited in *Memorial of Goode*, p. 405.

15. While recovering in Bermuda, Goode made clear his dissatisfaction in spending so much time on a relatively unimportant college museum. Goode to Baird, 6 March 1877, Baird MSS, RU 7002, SIA. *Catalogue of the Fishes of the Bermudas*, Smithsonian Institution, *Bulletin 5* (Washington, D.C., 1876).

16. The annual reports of the museum detail these acquisitions. Between 1883 and 1893, by Goode's estimation, museum holdings had expanded fifteen times and were, by the latter date, in excess of three million objects; see the U.S. National Museum, *Annual Report for 1891*, p. 9.

17. Quoted in Lane Cooper, *Louis Agassiz as a Teacher: Illustrative Extracts on His Method of Instruction* (Ithaca: Comstock Publishing Co., 1917), p. 39.

18. Goode, *Virginia Cousins. A Study of the Ancestry and Prosperity of John Goode of Whitby, A Virginia Colonial of the Seventeenth Century . . . 1148 to 1887* (Harrisburg, Va., C. J. Carrier Co., 1981; reprint of the 1887 edition).

19. Goode, "America's Relation to the Advance of Science," *Science*, n.s. 1 (1895), 8–9; Goode anticipated that in the future science might prove to be a conservative force.

20. Goode to Samuel H. Scudder, n.d., Scudder MSS, Boston Museum of Science, Boston, Massachusetts.

21. See the reprinted essays in this volume and also his compilation of documents and essays on *The Smithsonian Institution, 1846–1896: The History of Its First Half Century* (Washington, D.C., 1897).

22. "The Scientific Method of History," *Science*, 30 (May 1884): 564–5.

23. The relationship of Goode and the AHA is discussed at length in Kohlstedt, "History in a Natural History Museum," n. 1. After Goode's death, the AHA's relationship to the Smithsonian Institution gradually changed, as reflected in the correspondence of Samuel P. Langley to H. B. Adams and others, RU 31, Secretary, SIA.

24. On Peale's museum Charles Coleman Sellers, *Mr. Peale's Museum: Charles Willson Peale and the First Popular Museum of Natural Science and Art* (New York: Norton, 1980); for an example of a learned society see Simon Baatz, *Knowledge, Culture, and Science in the Metropolis: The New York Academy of Sciences, 1817–1970*, published in the *Annals of the New York Academy of Sciences*, vol. 584 (1990).

25. For a discussion of the British Museum's role, see Susan Sheets-Pyenson, *Cathedrals of Science: The Development of Colonial Natural History Museums during the*

Late Nineteenth Century (Kingston/Montreal: McGill-Queens University Press, 1988).

26. Barbara Kirshenblatt-Gimblett argues that Goode challenged an older antiquarian outlook by insisting that objects be textualized in her essay "Objects of Ethnology," to be published in *Exhibiting Culture: the Poetics and Politics of Museum Display* (Washington, D.C.: Smithsonian Institution Press, 1991).

27. Curtis M. Hinsley, Jr., *Savages and Scientists: The Smithsonian Institution and the Development of American Anthropology, 1846–1910* (Washington, D.C.: Smithsonian Institution Press, 1981).

28. *Classification of the Collection to Illustrate the Animal Resources of the United States. A List of Substances Derived from the Animal Kingdom, with Synopsis of the Useful and injurious Animals and a Classification of the Methods of Capture and Utilization,* published as a *Bulletin* of the United States National Museum (Washington, D.C., 1876).

29. Otis T. Mason, "The Educational Aspect of the United States," *Studies in Historical and Political Science,* 4 (1890), 511.

30. First published in the British Museum Association's *Annual Report* (1895), the essay was also separately published that same year in York, England, by Coultas and Volans; quotation on p. 3.

31. Goode, *Principles of Museum Administration,* pp. 28 and 37.

32. In Britain, Baird had a counterpart in William Henry Flower of the British Museum; Flower, with a number of other British scientists including Thomas Huxley, Alfred R. Wallace, and John Edward Gray, stressed the importance of having research collections separated from those for instruction. See, for example, Flower's *Essays on Museums and Other Subjects Connected with Natural History* (London: Macmillan and Co., 1898).

33. Goode began his essay "The Beginnings of Natural Science" with a direct quotation from the English philosopher whose work was widely read and cited in the United States. See, for example, Richard Hofstadter, *Social Darwinism in American Thought* (Boston: Beacon Press, 1955).

34. Goode also compiled a narrative history, "The Genesis of the U.S. National Museum," which appeared in the annual report of the National Museum for 1891 and was reprinted in *Memorial of Goode.*

35. *Memorial of Goode,* p. 19.

36. On this matter he had apparent disagreements with Samuel P. Langley, who followed Baird as Secretary of the Smithsonian and who wanted to supplement scientific displays with others that were ornamental and "interesting." See Langley to Goode, 19 December 1887, Assistant Secretary, RU 54, SIA.

37. G. Brown Goode to Herbert Baxter Adams, 12 December 1888, American Historical Association MSS, Library of Congress, Washington, D.C.

38. This statement, near the end of his last historical essay, concludes with the whimsical observation, "It may be that the use of the word naturalist is to become an anachronism."

39. "A Step Toward Scientific Self-Identity in the United States: The Failure of the National Institute," *Isis,* 62 (1971): 339–362.

40. This project is an outgrowth of my longstanding interest in Goode and has required extensive work in the Smithsonian Institution Archives. I have been assisted by nearly every member of that staff through the past decade and would like especially to acknowledge the help of William R. Massa, Jr., Susan L. McMurray, Alan L. Bain, Pamela Henson, and William Deiss. Without their help, this project might never have been completed.

BEGINNINGS OF NATURAL HISTORY IN AMERICA

1. Annual presidential address delivered at the sixth anniversary meeting of the Biological Society of Washington, February 6, 1886, in the lecture room of the United States National Museum.

2. James S. M. Anderson, History of the Church of England in the Colonies, p. 86, London, 1845–56.

3. Robert Hues, Tractatus de Globis, etc., 1611–63.

4. Harriot was also a friend and companion of Raleigh during his imprisonment in the Tower (1603–1616), and was his collaborator in the preparation of the History of the World. His fidelity was rewarded by that distinguished authority, Chief Justice Popham, who denounced him from the bench as "a devil."

5. Henry Hallam, Introduction to the Literature of Europe in the Fifteenth,

Sixteenth, and Seventeenth Centuries, 4th ed., 1854; I, pp. 454, 456; II, p. 223; III, p. 181. See also J. E. Montucla, Histoire des Mathematiques Ersch and Gruber, Algemeine Encyklopædie.

6. It would appear, however, that Wallis may have been too enthusiastic in his admiration of the English mathematician. Hallam states that he ascribed to Harriot a long list of discoveries which have since been reclaimed for Cardan and Vieta.

7. William Stith, History of The First Discovery and Settlement of Virginia, Williamsburg, 1747, p. 20.

8. John M. Good and Olinthus Gilbert Gregory, The Pantologia, V, 1813.

9. History of the First Discovery and Settlement of Virginia, Williamsburg, 1747, p. 20.

10. 1590. HARIOT (or Harriott), THOMAS. A Briefe and True Report | of the New Found Land of Virginia | of the commodities and of the nature and man | ners of the naturall inhabitants. Discouered by the English Colony there seated by Sir Richard | Greinuile Knight In the yeere 1585. Which rema | ined Vnder the gouernment of twelue monethes, | At the special charge and direction of the Honou- | rable SIR WALTER RALEIGH Knight lord Warden | of the stanneries Who therein hath beene fauoured | and authorised by her MAIESTIE | : and her letters patents: | This fore booke Is made in English | By Thomas Hariot, seruant to the above named | Sir WALTER, a member of the Colony and there | imployed in discouering | CUM GRATIA ET PRIVILEGIO CAES. MA^TISSPECIA^LI | Francoforti ad Mœnum | Typis Ioannis Wecheli, sumtibus vero Theodori | DeBry Anno CIC IC XC, | venales reperuunter in officina Sigismundi Feirabendii. | 4°. pp. 1–33 (1). Title page with ornamental border of architectural design.

11. There are now only six or seven perfect copies in existence. These, we are told by Sabine, are in the British Museum and Bodleian libraries, and in the private collections of Messrs. Lenox, Brown, Christie-Miller, and Mann, besides an imperfect copy in the library of Harvard College and one in the possession of Sir Thomas Phillipps. At a sale in London in 1883 a copy sold for £300. A reproduction in photolithographic facsimile was issued by Sabine in New York in 1875.

12. Edward Everett Hale's Life of Sir Ralph Lane. Archæologia Americana, IV, pp. 317–344.

13. Subsequently referred to by Champlain in 1613, and Sagard in 1636, under the name *chaousarou*, and figured by Champlain on his map of Nouvelle France. Du Creux, in his Historiæ Canadensis, 1664, also mentions it.

14. It has been generally supposed that Champlain was the first to notice this characteristic American animal, and Slafter, in his notes upon Champlain's works [Publications of the Prince Society, Champlain's Voyages, II, p. 87], makes a statement to that effect, and is followed by Higginson in his History of the United States. Actually, the French explorer did not observe it until twenty years after Harriot, and his account of it was not printed until 1613.

15. History of the First Discovery and Settlement of Virginia, Williamsburg, 1747, p. 16.

16. Sir Hans Sloane and additional Manuscripts, 5270.

17. Archæologia Americana, IV, pp. 21–24.

18. Sumario, Chap. XXVII, p. 491. Purchas, His Pilgrimmes, Chapter III, 1625, p. 995.

19. Idem, Chap. XI, p. 487.

20. A Paris edition of 1633 had the following title: Commentaire Royal ou l'Histoire des Yncas Roys de Peru, etc. Ensemble une description particulière des Animaux, des Fruits, des Mineraux, des Plantes, etc. Ecrite en langue Peruvienne et traduit fur la version Espagnole par I. Baudouin, Paris, 1633; Amsterdam, 1704 and 1715. See Artedi, Bibliotheca Ichthyologica, 1788, p. 65.

21. William Whewell, A History of the Inductive Sciences, from the Earliest to the Present Time, III, 1837, p. 325.

22. Carl Rau, Thiergärten. New Yorker Staats-Zeitung, April 26, 1863.

23. The golden eagle, says Aguilera.

24. Humming birds.

25. Trogons, known as *quetzales* by the Mexican Indians of to-day. Excellent examples of their pictorial use of trogon feathers may be seen in the United States National Museum.

26. *Cyanopiza versicolor.*

27. Ocelot, juguars, pumas, eyras, jaguarundis.

28. The coyote (*coyotl*), *Canis latrans*.

29. John G. Bourke, The Snake Dance of the Moquis of Arizona, New York, 1884.

30. Washington Matthews, Natural Naturalists. Bulletin of the Philosophical Society Washington, VII, 1885, p. 73 (abstract).

31. Timehri, being the Journal of the Royal Agricultural and Commercial Society of British Guiana. Demerara, I, 1882, pp. 25–43.

32. Generall Historie, 1624, p. 27.

33. From the Indian word *Moosoa*. Slafter, in his notes on Champlain's Voyages, I, p. 265, supposes the *Orignac* referred to by this explorer in his De Sauvages, etc., Paris, 1607, to have been the Moose, and his *Cerf* to have been the Caribou.

34. Generall Historie, 1624, pp. 216, 217.

35. A copy of this rare work was sold in London, 1883, for £69. A reprint was issued by Joel Munsell at Albany in 1860, but this privately printed edition consisted of only 200 copies and it is already scarce.

36. Page 21.

37. Doctor Cromwell Mortimer, in the dedication of Volume XL, Philosophical Transactions.

38. Tuckerman, in Archæologia Americana, IV, pp. 123–124. See also The Winthrop Papers. Massachusetts Historical Society Collections, 5th ser., VIII, p. 571.

39. Philosophical Transactions, XVII, pp. 781–795, 978–999; XVIII, pp. 121–135, and in Miscellaneoua Curiosa, III; also reprinted in Force's Historical Tracts, III.

40. Philosophical Transactions, XI, p. 6323.

41. Idem., XXI, p. 436.

42. Idem., XX, p. 167.

43. Calendar of Colonial Papers, IV, 1625, p. 77.

44. Idem., III, p. 75, Nos. 155, 156.

45. Calendar of Colonial Papers, I, 1638, p. 285.

46. Later known as Vauxhall Gardens, a famous place of resort.

47. The barber virtuoso, described in Bulwer's Devereux.

48. John Josselyn, An Account of Two Voyages to New England (made during the years 1638, 1663), Boston, 1865.

49. The Tempest, Act II, Scene 2.

50. In Nehemiah Grew's Catalogue and description of the natural and artificial Rarities, belonging to the Royal Society and preserved at Gresham College, Whereunto is subjoined the comparative Anatomy of Stomachs and Guts, London, 1694, are descriptions and figures of every American animal.

51. Archæologia Americana, IV, pp. 116, 117.

52. Publications of Prince Society, Boston, 1878; Hakluyt Society, XXIV, 1850.

53. Archæologia Americana, IV, p. 119.

54. Paris, 1672, octavo.

55. 1672, duodecimo, 2 vols.

56. Paris, 1698; Amsterdam, 1699; London (translation), 1698.

57. Paris, 1758.

58. Nouveaux Voyages aux Indes Occidentales, etc., Paris, 1768.

59. Nova Plantarum Ammericanarum Genera, 1793. Traité Des Fougères de L'Amérique, 1705.

60. First edition without name of author; others, Paris, 1665; Lyons, 1667; Amsterdam, 1716.

61. Brendel, American Naturalist, December, 1879, p. 757.

62. Philosophical Transactions, XVII, 1693, pp. 671, 672. See also Transactions of the Linnæan Society, VII, p. 227.

63. Some observations concerning Insects made by Mr. John Banister, in Virginia, A. D. 1680, with Remarks on them by Mr. James Petiver, etc. Philosophical Transactions, XXII, 1701, pp. 807–814.

64. Perhaps the *Megalonyx jeffersonii*, subsequently discovered.

65. In Ray's Historia Plantarum, London, 1686.

66. His papers and collections were sent to the Bishop of London. The plants are said to have passed into the hands of Sloane, and to be still preserved in the British Museum. It would be interesting to know what has become of his manuscripts.

67. John Lawson, History of North Carolina, Raleigh edition, p. 134.

68. See The Bland Papers and Slaughter's History of Bristol Parish, 1st and 2d editions.

69. Spotswood Letters, I, pp. 1, 8; II, pp. 44, 58, 354.

70. Darlington, Memorials of John Bartram and Humphrey Marshall, p. 21.

71. Dissertatio brevis de Principiis Botanicorum et Zoologorum, deque novo stabiliendo naturæ rerum congruo cum Appendice aliquot generum plantarum recens conditorum et in Virginia observatorum. Nuremburg, 1748.

72. Philosophical Transactions, XLIII, 1744, p. 102.

73. Idem., XLV, 1748, p. 541.

74. Idem., LI, Pt. 1759, p. 390.

75. American Medical and Philosophical Register, IV.

76. James Edward Smith, Correspondence of Linnæus, II, pp. 442–451.

77. James Thatcher, Medical Biography, I, P. 73.

78. Mitchell, writing to Linnæus, in 1748, remarks: " I can now only send you . . . some dissertations of Mr. Tennent upon the *Polygala*, two of which only have come out among his latest publications. His former ones, of inferior merit, are not now to be had."

79. Philosophical Transactions, XIX, 1697, p. 781.

80. Edward Tyson, Carigueya, seu Marsupialie Americanum, or the Anatomy of an Opossum, etc. Philosophical Transactions, XX, 1698, p. 105.

81. Brendel in American Naturalist, December, 1879, p. 756.

82. John Torrey, Flora of New York, Albany, 1843.

83. American Journal of Science, XLIV, 1843, p. 85.

84. London, 1754–1771.

85. Smith, Correspondence of Linnæus, I, p. 537.

86. William Darlington, Memorials of John Bartram and Humphrey Marshall. Philadelphia, 1849, 1850.

87. Charles C. Jones, History of Georgia. Boston and New York, 1863.

88. Darlington, Memorials of John Bartram and Humphrey Marshall. Philadelphia, 1849, 1850, p. 106.

89. Smith, Correspondence of Linnæus, I, p. 9.

90. Idem., p. 33.

91. Philosophical Transactions, XXXI, 1721, pp. 165–168.

92. Idem., XXXI, 1721, pp. 148–150.

93. Idem., XXXII, 1723, pp. 292–295.

94. Idem., XXXIII, 1725, pp. 256–269.

95. Idem., XXI, 1721, pp. 145, 146.

96. Idem., XXXIII, 1724, pp. 194–200.

97. See Tuckerman in Archæologia Americana, IV, pp. 125, 126.

98. Magazine of American History, April, 1885, p. 379.

99. Magazine of American History, April, 1885, p. 386.

100. Erlangen, 1788, 2 vols., octavo.

101. Philosophical Transactions, L, Pt. 2, 1758, p. 859.

102. Smith, Correspondence of Linnæus, II, pp. 507–550.

103. See similar speculations in George Scot's Model of the Government of the Province of East New Jersey in America. Edinburgh, 1685.

104. Memoirs of the American Academy of Sciences, 1785.

105. Brendel, American Naturalist, December, 1879, p. 758.

106. John Eliot, Biographical Dictionary of Eminent Characters in New England. Boston, 1809.

107. See Dossie, Memoirs of Agriculture. London, I, 1768, pp. 24–26; also Brock in Richmond Standard, April 26, 1879, p. 4.

108. Smith, correspondence of Linnæus, I, p. 477.

109. Samuel Mordecai, Richmond in By-gone days. Richmond, 1856. A copy of the original pamphlet of proposals is still preserved in the Virginia State Library.

110. One of the original tickets to these courses is in the Library of the Surgeon-General's Office in Washington.

111. Darlington, Memorials of John Bartram and Humphrey Marshall, p. 535.

112. Seaside Studies in Natural History, p. 25.

113. The genus *Harriotta* has been dedicated by Goode and Bean to the memory of Thomas Harriot. It is intended to embrace a long-rostrated chimæroid fish from deep water off the Atlantic coast of North America. The description is not yet published. Heriot's Isle, named for Harriot by the early explorers, and shown upon Vaughan's map, in Smith's General History of Virginia, has entirely disappeared. It was situate on the north side of Albemarle Sound, about midway

between Roanoke Island and the mouth of Chowan River. Whether it has been swept away by tides, or has become a part of the mainland, it is difficult to say. The latter supposition seems the most probable, and since it is in all likelihood Reeds Point which now occupies its former location, the propriety is suggested of calling this little cape Harriots Point, in memory of the explorer.

BEGINNINGS OF AMERICAN SCIENCE

1. Annual presidential address delivered at the seventh anniversary meeting of the Biological Society of Washington, January 22, 1887, in the lecture room of the United States National Museum.

2. See obituary in the European Magazine, July, 1796; also Memoirs of Rittenhouse, by William Barton, 1813, and Eulogium by Benjamin Rush, 1796.

3. Von Zach, Monatliche Correspondenz, II, p. 215.

4. Philadelphia Medical and Physical Journal, I, Pt. 2, p. 96.

5. Barton's Memoirs of Rittenhouse, 1813, p. 614.

6. James Thacher, American Medical Biography, I, 1828, p. 408.

7. This eventually became the property of the university. See Barton's Memoirs of Rittenhouse, 1813, p. 377. Transactions of the American Philosophical Society, II, p. 368.

8. David Hosack, Tribute to the Memory of the late Caspar Wistar, New York, 1818.

9. Caleb Atwater, Remarks made on a Tour to Prairie du Chien; thence to Washington City, in 1829. Columbus, Ohio, 1831, p. 238.

10. Standard Natural History, pp. lxii–lxxii.

11. See previous address, p. 99. [This volume, p. 83.]

12. Benjamin S. Barton, Transactions of the American Philosophical Society, III, p. 339.

13. I often heard the great Linnæus wish that he could have explored the continent of North America. Collin, Transactions of the American Philosophical Society, III, p. xv.

14. Idem., p. xxiv.

15. Thomas Godfrey [says a recent authority] was born in Bristol, Pennsylvania, in 1704, and died in Philadelphia in December, 1749. He followed the trade of a glazier in the metropolis, and, having a fondness for mathematical studies, marked such books as he met with, subsequently acquiring Latin, that he might become familiar with the mathematical work in that language. Having obtained a copy of Newton's Principia, he describes an improvement he had made in Davis' quadrant to James Logan, who was so impressed that he at once addressed a letter to Edmund Halley in England, giving a full description of the construction and uses of Godfrey's instrument.

16. Benjamin Silliman, Jr., American Contributions to Chemistry, p. 13.

17. See Priestley's History of Electricity.

18. See previous address, p. 99. [This volume, p. 82]

19. Cyclopædia of American Biography, III, p. 260.

20. Elliott Coues, Key to North American Birds, 1887, p. xvii.

21. Transactions of the American Ethnological Society, III, 1851.

22. Nicholson's Journal, 1805.

23. Travels through North and South Carolina, Georgia, East and West Florida, 1794, p. xiii.

24. William J. Hooker, On the Botany of America, Edinburgh Journal of Science, II, p. 108.

25. William P. C. Barton, Biography of Benjamin S. Barton, Philadelphia, 1815.

26. David Starr Jordan, Bulletin U. S. National Museum, No. IX; also see article in Popular Science Monthly, XXIX, p. 212, reprinted in Jordan's Science Sketches, p. 143.

27. Chase, Potter's American Monthly, VI, pp. 97–101.

28. Audubon, The Eccentric Naturalist, Ornithological Biography, p. 455.

29. American Monthly Magazine, II, p. 81.

30. Idem, I, p. 80.

31. Notes on the State of Virginia, 1788, pp. 69–71.

32. Examples of his verses may be found in Duyckinck's Cyclopædia of American Literature, I, p. 520.

33. See John W. Francis, Life of Doctor Mitchill, in Williams's American

Medical Biography, pp. 401–411, and eulogy in Discourse in Commemoration of Fifty-third Anniversary of the New York Historical Society, 1857, pp. 56–60; and in his Old New York; also—

Sketch by H. L. Fairchild, in History of the New York Academy of Sciences, 1887, pp. 57–67; also Doctor Mitchill's own pamphlet: Some of the Memorable Events and Occurrences in the Life of Samuel L. Mitchill, of New York, from the year 1786 to 1827.

A biography by Akerly was in existence, but has never been printed.

Numerous portraits are in existence, which are described by Fairchild.

34. David Hosack, Memoirs of De Witt Clinton, New York, 1829. James Renwick, Life of De Witt Clinton, New York 1840. William W. Campbell, Life and Writings of De Witt Clinton, New York, 1849.

35. John H. Griscom, Memoir of John Griscom, New York, 1859, p. 424.

36. De Witt Clinton, Transactions of the Literary and Philosophical Society, New York, I, p. 59.

37. John T. Kirkland, Memoirs of the American Academy, New Series, I, p. xxii.

38. See previous address, p. 95. [This volume, p. 379]

39. Memoirs of the American Academy of Sciences, II, Pt. 2, 1797, p. 46.

40. Biography in Polyanthus, II.

41. Walpole, New Hampshire, 1794, 8vo, p. 416.

42. See previous discourse, p. 98. [This volume, p. 82.]

43. Charles Darwin, Origin of Species, 6th Amer. ed., p. xv. Edward S. Morse, Proceedings of the American Association for the Advancement of Science, XXV, p. 141.

44. Erasmus, grandfather of Charles Darwin.

45. There is a whole series of quarto or folio volumes in the British Museum done by him, and a few volumes are extant in this country. Besides, all the biological material in Smith-Abbot's Insects of Georgia is his.—Letter of S. H. Scudder.

46. Transaction of the American Philosophical Society, I, 1789, p. 274.

47. Idem, 1798, p. 294.

48. The first vertebrate fossils were found in Virginia. Samuel Maverick, of

Massachusetts, reported to the colony at Boston in 1636 that, at a place on the James River, about 60 miles above its mouth, the colonists had found shells and bones, among these bones that of a whale 18 feet below the surface.—Neill's Virginia Carolorum, p. 131.

49. Histoire des Chènes de l'Amerique Septentrionale, 1801; 36 plates.

50. Voyage à l'ouest des Monts Alléghanys, etc., octavo, pp. 684. Paris, 1808. Histoire des Arbres Foréstières de l'Amerique Septentrionale.

51. Viaggio negli Stati Uniti del l'America Settentrionale.

52. Tableau du climat et du sol des Etats-Unis d'Amérique, suivi d'éclaircissements sur la Floride, sur la colonie française à Scioto sur quelques colonies canadiennes, et sur les sauvages. Paris, 1803. Octavo, 2 vols. 2d edition. Paris. Octavo, 1 vol., pp. 494. Map.

53. John Bigelow, Franklin's Home and Host in France. The Century Magazine, May, 1888, p. 743.

54. See a complete bibliography of the various reports of this expedition, by Elliott Coues, in the Bulletin of the United States Geological Survey.

55. See American Journal of Science, XLII, 1842, p. 5.

56. Jefferson's Works (edited by T. J. Randolph), 1830, III, p. 461.

57. American Journal of Science, I, 1819, 37.

58. T. H. Wood, The Reign of Victoria; a survey of Fifty Years of Progress. London, 1887.

59. Geological Text-Book, 2d edition, 1832, p. 16.

60. The Troy Lyceum of Natural History was incorporated in 1819, and a lectureship was created, filed by Mr. Eaton (American Journal of Science, II, p. 173). In 1820 a similar association, The Hudson Association for Improvement in Science, was founded in the city of Hudson, and in 1821 the Delaware Chemical and Geological Society.

61. Proceedings of the American Association for the Advancement of Science, VI, 1851, pp. vi, xlvi.

62. American Journal of Science, III, 1821, pp. 201–216.

63. Martin, Memoir of William Maclure, p. 11.

64. Index to the Geology of the Northern States, 2d edition, 1820, p. viii.

65. "No future historian of American science will fail to commemorate this

work as our earliest *purely scientific* Journal, supported by *original American communications*," said Silliman in his prospectus, 1817.

66. Baltimore has a handsome museum, superintended by one of the Peale family, well known for their devotion to natural science and to works of art. It is not their fault if the specimens which they are enabled to display in the latter department are very inferior to their splendid exhibitions in the former.—Mrs. Trollope, Domestic Manners of the Americans. London, I, 1832, p. 296.

67. Transactions of the American Philosophical Society, II, p. 366.

68. An address to the people of the Western Country, dated Cincinnati, September 15, 1818, and signed by Elijah Slack, James Findlay, William Steele, Jesse Embrees, and Daniel Drake, managers.

69. Histoire Générale et Iconographie des Lépidoptères et des Chenilles de l'Amérique Septentrionale, Paris, 1830.

70. Loudon's Gardeners' Magazine.

71. A. H. Stephens in Johnson's New Universal Cyclopædia, New York, 1876, II, p. 1702.

72. The LeConte family deserves a place in Galton's Hereditary Genius. Professor John LeConte, the physicist, and Professor Joseph LeConte, the geologist, were sons of Doctor Lewis LeConte, while Doctor J. L. LeConte was the son of Major John Eatton LeConte.

73. Contributions of the Maclurian Lyceum, I, January, 1827, p. 3.

74. Idem., I, p. 41.

75. American Journal of Science, X, 1826, p. 369.

76. See History of the New York Academy of Sciences, 1887, p. 76.

77. See Rev. John McIlwraith's Life of Sir John Richardson, C. B., LL. D. London, 1868. Also Obituary in London Reader, 1865, p. 707.

78. See a Memoir by B. H. Coates, read before American Philosophical Society, December 16, 1834; a memoir by George Ord; also a tribute to his memory in Dall's presidential address before the Biological Society of Washington in January, 1888.

79. George Ord, Memoir of Charles Alexander Lesueur. American Journal of Science, VIII, 1849, p. 189.

80. Voyage des Decouvertes aux Terres Australes.

81. Voyage aux Terres Australes, Paris, 1807.

82. Volume I, p. 439.

83. David Brewster, Edinburgh Journal of Science, II, 1825, p. 108.

84. Theodore Gill.

85. Reviewed in American Journal of Science, III, 1821, p. 47, and in Blackwood's Magazine, XVI, 1824, p. 420; XVII, 1825, p. 56.

86. American Medical Recorder, VII, 1824, p. 223.

87. American Journal of Science, IV, 1822, p. 283.

88. Idem, XXIV, 1833, p. 349.

89. Journal of the American Silk Company, I, May, 1839, p. 179.

90. Proceedings of the American Association for the Advancement of Science, II, 1849, p. 163.

91. B. A. Gould, Address in Commemoration of Sears Cook Walker. Proceedings of the American Association for the Advancement of Science, VIII, 1854, p. 25.

92. Benjamin Peirce, Cambridge Miscellany, April, 1842, p. 25.

93. Proceedings of the American Association for the Advancement of Science, 1849, II, p. 164.

94. Proceedings of the American Association for the Advancement of Science, VI, p. xlviii.

95. See Andrew D. White's Scientific and Industrial Education in the United States. Popular Science Monthly, V, 1874, p. 170.

96. North American Review, January, 1876, pp. 100, 107.

ORIGIN OF NATIONAL SCIENTIFIC AND EDUCATIONAL INSTITUTIONS

1. A paper presented before the American Historical Society at the meeting held in Washington in 1889, and revised and corrected by the author to July 15, 1890.

2. John Winthrop, F.R.S. [1606–1676], elected governor of Connecticut in 1657.

3. John Eliot, Biographical Dictionary of Eminent Characters in New England. Boston, 1809.

4. The first meetings of the body of men afterwards organized as the Royal Society appear to have taken place during the Revolution and in the time of Cromwell; and as early as 1645, we are told by Wallace, weekly meetings were held of "divers worthy persons inquisitive into natural philosophy and other parts of human learning, and particularly of what has been called the new philosophy, or experimental philosophy," and it is more than probable that this assembly of philosophers was identical with the Invisible College, of which Boyle spoke in sundry letters written in 1646 and 1647. These meetings continued to be held, sometimes at the Bull Head Tavern, in Cheapside, but more frequently at Gresham College, until 1660, when the first record book of this society was opened. Among the first entries is a reference to a design then entertained "of founding a college for the promoting of physico-mathematicall experimentall learning." Doctor Wilkins was appointed chairman of the society, and shortly after, the King, Charles II, having become a member, its regular meeting place was appointed to be in Gresham College.

5. This name was adopted in 1768 to replace that first adopted in 1766, which was The American Society for Promoting and Propagating Useful Knowledge, held in Philadelphia.

6. Some insight into the scientific politics of the time may be gained by reading the following extract from a letter addressed to Franklin by Doctor Thomas Bond, June 7, 1769: I long meditated a revival of our American Philosophical Society, and at length thought I saw my way clear in doing it; but the old party leaven split us for a time. We are now united, and with your presence may make a figure; but till that happy event I fear much will not be done. The assembly have countenanced and encouraged us generously and kindly; and we are much obliged to you for your care in procuring the telescope which was used in the late observations of the transit of Venus.

7. A copy of the finished volume of the Transactions was presented to each member of the Pennsylvania assembly, accompanied by an address as follows: As the various societies which have of late years been instituted in Europe have confessedly contributed much to the more general propagation of knowledge

and useful arts, it is hoped it will give satisfaction to the members of the honorable house to find that the province which they represent can boast of the first society and the first publication of a volume of Transactions for the advancement of the useful knowledge of this side of the Atlantic—a volume which is wholly American in composition, printing, and paper, and which, we flatter ourselves, may not be thought altogether unworthy of the attention of men of letters in the most improved parts of the world.

8. Some local antiquary may make an interesting contribution to the literature of American museum work by looking up the history of this collection.

9. The Chevalier Anne César de la Luzerne (1741–1821] was French Minister to the United States from 1779 to 1783; afterwards minister to England. M. François de Barbé Marbois [1745–1837] was his secretary of legation, and after the return of his chief to France, was chargé d'affaires until 1785. For many interesting facts, not elsewhere accessible, concerning the career of these men in the United States, and their acquaintance with Adams, see John Durand's admirable New Materials for a History of the American Revolution. New York: Henry Holt & Co., 1889, 12mo, pp. i–vi, 1–310.

10. Rev. Samuel Cooper, D. D. [1725–1783], an eminent patriot, long pastor of Brattle Street Church, in Boston, and a leading member of the corporation of Harvard. He was the first vice-president of the American Academy of Arts and Sciences.

The first president of the academy was James Bowdoin, afterwards governor of Massachusetts, and the friend of Washington and Franklin, and a member of the Royal Society. He held the presidency from 1780 until his death in 1790. His descendant, the Hon. Robert C. Winthrop, was chosen to deliver the oration at the centennial anniversary of the organization of the society.

11. The provision in the State constitution of which Mr. Adams speaks, was the following:

The encouragement of literature, etc. Wisdom and knowledge, as well as virtue diffused generally among the body of the people, being necessary for the preservation of their rights and liberties, and as these depend on spreading the opportunities and advantages of education in the various parts of the country, and among the different orders of the people, it shall be the duty of legislators and magistrates in all future periods of the Commonwealth, to cherish the interests of literature and the sciences, and all seminaries of them, especially the university at Cambridge, public schools, and grammar schools in the towns, to encourage private societies and public institutions, rewards and immunities for the promotion of agriculture, arts, sciences, commerce, trades, manufactures, and a natural history of the country; to countenance and inculcate the principles of humanity and general benevolence, public and private charity, industry and frugality, honesty and punctuality in their dealings, sincerity, good humor, and all social affections and generous sentiments among the people.

This feature of the constitution of Massachusetts, [writes Mr. Adam's biographer,] is peculiar, and in one sense original with Mr. Adams. The recognition of the obligation of a State to promote a higher and more extended policy than is embraced in the protection of the temporal interests and political rights of the individual, however understood among enlightened minds, had not at that time been formally made a part of the organic law. Those clauses since inserted in other State constitutions, which, with more or less of fullness, acknowledged the same principle, are all manifestly taken from this source.

12. Letter of Manasseh Cutler to Doctor Jonathan Stokes, August 17, 1785.

13. Idem.

14. Copies of Quesnay's pamphlet are preserved in the Virginia State library at Richmond and in the Andrew D. White Historical library of Cornell University, as well as in a certain private library in Baltimore. A full account of this enterprise may be found in Herbert B. Adams's Thomas Jefferson and the University of Virginia, pp. 21–30, and other records occur in Mordecai's Richmond in By-gone Days (2d edition, pp. 198–208) and in Goode's Virginia Cousins, p. 57.

The building erected for the Academy of Sciences was the meeting place of the convention of patriots and statesmen who ratified in 1788 the Constitution of the United States, and subsequently was the principal theater of the city of Richmond.

The academy grounds, [writes R. A. Brock,] included the square bounded by Broad and Marshall and Eleventh and Twelfth streets, on the lower portion of which stood the Monumental Church and the medical college. The academy stood midway in the square fronting Board street. L'Académie des Etats-Unis de l'Amérique was an attempt, growing out of the French alliance with the United States, to plant in Richmond a kind of French academy of the arts and sciences, with branch academies in Baltimore, Philadelphia, and New York. The institution was to be at once national and international. It was to be affiliated with the royal societies of London, Paris, Brussels, and other learned bodies in Europe. It was to be composed of a president, vice-president, six counsellors, a treasurer-general, a secretary, and a recorder, an agent for taking European subscriptions, French professors, masters, artists in chief attached to the academy, 25 resident and 175 nonresident associates, selected from the best talent of the Old World and the New. The academy proposed to publish yearly, from its own press in Paris, an almanac. The academy was to show its zeal for science by communicating to France and other European countries a knowledge of the natural products of North America. The museums and cabinets of the Old World were to be enriched by the specimens of the flora and fauna of a country as yet undiscovered by men of science. The promoter of this brilliant scheme was the Chevalier Alexander Maria Quesnay de Beaurepaire, grandson of the famous French philosopher and economist, Doctor Quesnay, who was the court physician of Louis XV. Chevalier Quesnay had served as a captain in Virginia, in 1777–78, in the war of the Revolution. The idea of founding the academy was suggested to him in 1778 by John Page, of Rosewell, then lieutenant governor of Virginia, and himself devoted to scientific investigation. Quesnay succeeded in raising by subscription the sum of 60,000 francs, the subscribers in Virginia embracing nearly 100 prominent names. The corner stone of the building, which was of wood, was laid with Masonic ceremonies July 8, 1786. Having founded and organized this academy under the most distinguished auspices, Quesnay returned to Paris and succeeded in enlisting in support of his plan many learned and distinguished men of France and England. The French revolution, however, put an end to the scheme. The academy building was early converted into a theater, which was destroyed by fire, but a new theater was erected in the rear of the old. This new building was also destroyed by fire on the night of December 26, 1811, when 72 persons perished in the flames. The Monumental church commemorates the disaster, and its portico covers the tomb and ashes of most of its victims. A valuable sketch of Quesnay's enlightened projection, chiefly dawn from his curious Mémoire concernant l'Académie des Sciences et Beaux-Arts des Etats-Unis d'Amérique, établie à Richmond, was published in The Academy, December, 1887, II, No. 9, pp. 403, 412, by Doctor Herbert B. Adams, of Johns Hopkins University. A copy of Quesnay's rare Mémoire is in the library of the State of Virginia. Quesnay complains bitterly that all his letters relating to his service in the American Army had been stolen from a pigeonhole in Governor Henry's desk and his promotion thus prevented.

15. In an article recently published by Professor C. V. Riley, he sustains the popular belief and tradition that *Cecidomya* was introduced about the time of the Revolution, and probably by Hessian troops. He gives interesting details concerning the work of the committee of the American Philosophical Society, and a review of recent controversies upon this subject. See Canadian Entomologist, XX, p. 121.

16. Before the organization of the Department of Agriculture, another step in economic entomology was taken by the General Government in the publication of an official document on silk worms:

1828. | Mease, James. | 20th Congress, | 18th Session | [Doc. No. 226] Ho. of Reps. | Silk-worms | ——— | Letter | from | James Mease, | transmitting a treatise on the rearing of silk-worms, | by Mr. De Hozze, of Munich, | with plates, etc., etc. | ——— | February 2, 1828.—Read and referred to the Committee on Agriculture. | ——— | Washington: | Printed by Gales and Seaton | 1828. | 8°, pp. 1–108.

17. Probably the third William Byrd [1728–1777], the son of the author of Westover Papers. He was colonel of the Second Virginia Regiment in 1756, and perhaps was in camp with Washington on the present site of the capital, when he became so deeply impressed with the eligibility of the site for a national city.

18. Perhaps Lewis Evans, the geographer, who in 1749 published a map of the central colonies, including Virginia. Professor Winsor tells me that there are copies of this map in the possession of Harvard University, in the library of the Pennsylvania Historical Society, and one in the Faden collection in the Library of Congress. Professor Josiah D. Whitney says that the legend on it, "All great storms begin to leeward," is, so far as he knows, the first expression of that scientific opinion.

19. Economica, p. 22.

20. Idem, Appendix, p. ix.

21. Madison Papers, I, pp. 354, 577.

22. See Appendix A.

23. See Appendix B.

24. The Society of Sons of the American Revolution, recently organized, and composed of descendants of Revolutionary soldiers and patriots, has for one of its objects "to carry out Washington's injunction 'to promote as objects of primary importance institutions for the diffusion of knowledge,' and thus to create an enlightened public opinion."

25. 1806 Blodget, Samuel, jr., Economica: | A Statistical Manual | for the | United States of America. | = | .... The legislature ought to make the people happy | Aristotle on government | = | .... "Felis qui potuit rerum cognoscere causas" | = | City of Washington: | Printed for the author. | = | 1806, 128 i–viii, 1–202 i–xiv.

The certificate of copyright is in this form:

Be it remembered that * * * Samuel Blodget, junior, hath deposited in this office, the title of a book the right whereof he claims as author, but for the benefit in trust for the free education fund of the university founded by George Washington in his last will, etc.

26. At least three fascicles of Extracts from the minutes of the United States Military Society were printed—one for the stated meeting, October 6, 1806 [4°, 14 pp.]; one for an occasional meeting at Washington, January 30, 1808 [4°, pp. 1–23 (l)]; and one for an occasional meeting at New York, December 28, 1809 [4°, pp. 1–22]. The manuscript records, in four volumes, are said to be in the possession of the New York Historical Society.

I am indebted to Colonel John M. Wilson, United States Army, Superintendent of the Military Academy, and to General J. C. Kelton, United States Army, for courteous and valuable replies to my letters of inquiry.

27. Statutes at Large, I, pp. 109–112.

28. Among the treasures of the National Museum is a patent dated 1796, signed by Washington as President and Pickering as Secretary of State.

29. The foregoing paragraphs concerning the history of the Patent Office were kindly supplied by Mr. Edward Farquhar, for many years its assistant librarian.

30. See Official Gazette, United States Patent Office, XII, No. 15, Tuesday, October 9, 1877; also articles in Appleton's and Johnson's Cyclopædias.

The history of the Patent Office has never been written; a full account of its work and of its influence upon the progress of American invention is greatly to be desired.

31. See Jefferson, A Memoir on the Discovery of Certain Bones of a Quadruped, of the Clawed kind, in the Western Part of Virginia, in the American Philosophical Transactions, IV, p. 246 (March 10, 1797); also F. B. Luther, Jefferson as a Naturalist, in the Magazine of American History, April, 1885, pp. 379–390.

32. Go, wretch, resign the Presidential chair;
   Disclose thy secret measures, foul or fair.
   Go, search with curious eyes for hornèd frogs
   'Mid the wild wastes of Louisianian bogs,
   Or where the Ohio rolls his turbid stream
   Dig for huge bones, thy glory and thy theme.

33. Todd, Life and Letters of Joel Barlow, p. 208.

34. Adams, Jefferson and the University of Virginia, p. 49 *et seq.*

35. See text of prospectus in Appendix C to this paper, or in National Intelligencer, Washington, 1806, August 1 and November 24. The original publication, of which there is a copy in the Congressional Library, recently brought to my notice by Mr. Spofford, is a pamphlet, anonymously published, with the date of Washington, 24th January, 1806.

Prospectus | of a | National Institution, | to be | established | in the | United States | ═ | Washington City: | Printed by Samuel H. Smith |———| 1806— 8°, pp. 1–44.

36. National Intelligencer, November 24, 1806.

37. Henry Adams, History of the United States, 1805–1809, I, pp. 346, 347; II, p. 365.

38. Idem., pp. 45, 46.

39. The Old Bachelor, by William Wirt, p. 186.

40. I am indebted to Doctor James C. Welling, president of the Columbia University, for much important information concerning this and other matters discussed in the present paper.

41. James C. Welling, The Columbian University, Washington, 1889, p. I. The following letter, written by President Monroe in 1821, indicates that the public men of the day were not unwilling that the institution should be regarded as one of national scope:

WASHINGTON, *March 28, 1821.*

SIR: I avail myself of this mode of assuring you of my earnest desire that the college which was incorporated by an act of Congress at the last session, by the title of The Columbian College in the District of Columbia, may accomplish all the useful purposes for which it was established; and I add, with great satisfaction, that there is good reason to believe that the hopes of those who have so patriotically contributed to advance it to its present stage will not be disappointed.

Its commencement will be under circumstances very favorable to its success. * * * The act of incorporation is will digested, looks to the proper objects, and grants the powers well adapted to their attainment. The establishment of the institution within the Federal District, in the presence of Congress, and of all the departments of the Government, will secure to the young men who may be educated in it many important advantages; among which, the opportunity which it will afford them of hearing the debates in Congress, and in the Supreme Court, on important subjects, must be obvious to all.

With these peculiar advantages, this institution, if it receives hereafter the proper encouragement, *cannot fail to be eminently useful to the nation.* Under this impression, I trust that such encouragement will not be withheld from it.

I am, sir, with great respect, your very obedient servant,

JAMES MONROE

42. Life of Josiah Meigs, p. 102.

43. The original members of the Columbian Institute were: Hon. John Quincy Adams; Colonel George Bomford, U. S. A.; Doctor John A. Brereton, U. S. A.; Doctor Edward Cutbush, U. S. N.; Asbury Dickins, esq.; Joseph Gales, jr., esq.; Doctor Henry Huntt; Thomas Law, esq.; Edmund Law, esq.; Doctor George W. May; Alexander McWilliams, esq.; William Winston Seaton, esq.; Samuel H. Smith, esq.; William Thornton, esq.; Hon. Roger C. Weightman.

Among the later members were Doctor Joseph Lovell, U. S. A.; Colonel Isaac Roberdeau; Doctor Thomas Sewell; Judge William Cranch; Hon. Henry Clay; Hon. John McLean; Hon. Richard Rush; Hon. S. L. Southard; Hon. William Wirt; Doctor W. S. W. Ruschenberger, U. S. N.; Hon. J. M. Berrien; Hon. John C. Calhoun; Rev. Obadiah B. Brown, and Rev. William Staughton.

The minutes of the Columbian Institute are not to be found. The treasurer's book is in the National Museum.

44. This appropriation was made on the strength of a report by Senator Barbour, of Virginia, chairman of the Committee on the District of Columbia, in which, after alluding to the long-recognized "utility of a central literary establishment" and to the failures of the recomnmendations of Washington and Madison, he gave a brief history of the enterprise, which was as follows:

At length a few enterprising and patriotic individuals attempted to achieve, by voluntary donations, that which it had been supposed could be effected only by the power of Congress.

Their efforts were crowned with distinguished success. One individual in particular, the Rev. Luther Rice, with an unwearied industry and an unyielding perseverance which prompted him to traverse every part of the Union in pursuit of aid to this beneficent object, contributed principally to that success.

The funds thus acquired were faithfully and judiciously applied to the object. * * * Application was made to Congress for an act of incorporation, which passed February 9, 1821. This, however, was all the aid which Congress dispensed.

The accompanying document shows that there have been expended on this institution $80,000, $50,000 only have been procured; and, as a consequence, this institution is embarrassed with a debt to the amount of $30,000. * * * Under these circumstances, the individuals who have thus generously devoted themselves to the promotion of this establishment, and who have disinterestedly pledged their independence upon the successs of the college, present themselves to Congress, with a view to obtain their protection by a small pecuniary grant.

The committee, in reviewing the peculiar circumstances which characterize the origin of this establishment, its progress, and the great benefits it promises to society, are of opinion that the application is reasonable. It cannot be doubted, had such an establishment grown up, under similar circumstances, in either of the states, it would receive the helping hand of its Legislature. Congress stands in the same relation to this establishment, from its exclusive power of legislation within the District.

Report of Mr. Barbour, from the Committee on the District of Columbia, to whom was referred the memorial of the trustees of the Columbian College. April 19, 1824. Senate, Eighteenth Congress, first session (67). pp. 80–83.

45. The only suggestion which has ever been offered is that by Mr. W. J. Rhees, in his history of James Smithson and his Bequest, in which he calls attention to the fact that in the library of Smithson was a copy of Travels through North America, published in 1807 by Isaac Weld, secretary of the Royal Society, in which he describes the city of Washington, and refers to it prophetically as likely some time to become the intellectual and political center of one of the greatest nations of the world.

46. The Old Bachelor, p. 171. Baltimore: F. Lucas, jr. Small 8vo, pp. 1–235.

47. Life of Matthew Fontaine Maury," by Mrs. D. F. M. Corbin, London, 1888, p. 6.

48. Jefferson does not mention in this connection the well-known fact that he himself became personally responsible for raising the sum of 1,000 guineas from private sources to secure the sending out of this expedition.

49. The late Doctor Asa Gray, in a letter written to me shortly before his death, remarks: "I have reason to think that Michaux suggested to Jefferson the expedition which the latter was active in sending over to the Pacific. I wonder if he put off Michaux for the sake of having it in American hands."

I think it is sufficiently evident from what has been written, that the project had been considered by Jefferson long before Michaux came into America. A statement parallel to that of Jefferson is found in the brief biography of Michaux prefixed by Professor C. S. Sargent, to his reprint of the Journal of André Michaux, published in the Proceedings of the American Philosophical Society, XXVI, No. 129, p. 4: The French government was anxious at this time to introduce into the royal plantations the most valuable trees of eastern North America, and Michaux was selected for this undertaking. He was instructed to explore the territory of the United States, to gather seeds of trees, shrubs, and other plants, and to establish a nursery near New York for their reception, and afterwards to send them to France, where they were to be planted in the Park of Rambouillet. He was directed also to send game birds from America with a view to their introduction into the plantations of American trees. Michaux, accompanied by his son, then fifteen years old, arrived in New York in October, 1785. Here, during two years, he made his principal residence, establishing a nursery, of which all trace has now disappeared, and making a number of short botanical journeys into New Jersey, Pennsylvania, and Maryland. The fruits of these preliminary explorations, including twelve boxes of seeds, five thousand seedling trees, and a number of live partridges, were sent to Paris at the end of the first year.

Michaux's first visit to South Carolina was made in September, 1787. He found Charleston a more suitable place for his nurseries, and made that city his headquarters during the rest of his stay in America.

Michaux's journeys in this country after his establishment in Charleston are detailed in the Journal [printed in the place already referred to]. They cover the territory of North America from Hudson's Bay to the Indian river in Florida, and from the Bahama islands to the banks of the Mississippi river. His ambition to carry out his instructions was equaled only by his courage and industry. The history of botanical exploration records no greater display of fortitude and enthusiasm in the pursuit of knowledge, than Michaux showed in his journey to the headwaters of the Savannah river in December, 1788, when his zeal was rewarded by the discovery of *Shortia* or in the return from his visit to Hudson's Bay. The hardship of his last journey even did not satisfy his cravings for adventure and discovery; and shortly after his return he laid before the American Philosophcal Society a proposition to explore the unknown region which extended beyond the Missouri. His proposition was well received. The sum of five thousand dollars was raised by subscription to meet the expenses of the journey; all arrrangements were made and he was about to start when he was called upon by the Minister of the French Republic, lately arrived in New York, to proceed to Kentucky, to execute some business growing out of the relations between France and Spain with regard to the transfer of Louisiana.

It was this suggestion of Michaux, no doubt, [says Sargent in concluding this reference,] which led Mr. Jefferson, who had regarded it with great favor, to send a few years later the first transcontinental expedition to the shores of the Pacific.

Professor Sargent, like Doctor Gray, has evidently not been in possession of the history of Jefferson's early interest in this matter.

50. Jefferson's Writings, ed. T. J. Randolph, IV, pp. 513, 514.

51. It is a matter of history that Alexander Wilson, the ornithologist, was anxious to be appointed the naturalist of Pike's expedition, and Jefferson has been warmly abused for not gratifying his desire. It should be borne in mind that at this time Wilson was a man whose reputation had not yet been achieved, and also that it is quite possible that in those days, as in the present, the projectors of such enterprises were often hindered by lack of financial opportunity.

52. The United States Geological Survey was organized March 3, 1879, and Clarence King was appointed its first director. Major J. W. Powell, his successor, was appointed March 18, 1881.

53. The committee of twenty, appointed in 1857 by the American Association for the Advancement of Science to report upon the history and progress of the Coast Survey, made the following statement:

It is believed that the honor of first suggesting a geodetic survey of the American coast, is due to the elder Professor Patterson, of Philadelphia; who, as early as the year 1806, availed himself of his intimacy with the President, Mr. Jefferson, and the gentlemen who formed his cabinet, to impress them with the feasibility and policy of the measure. (Report on the History and Progress of the American Coast Survey up to the year 1858, by the Committee of Twenty, appointed by the Association for the Advancement of Science, at the Montreal meeting, August, 1857 (pp. 1–126), p. 23.)

54. I arrived in this country in October, 1805, having relinquished my public station in my native country, Switzerland, foreseeing the turn of political events which have since come to pass, and from a taste for a rural life with completely different views and means quite sufficient for them, but which I have failed to claim. Having arrived in Philadelphia, the late Professor Patterson, Mr. Garnet, of New Brunswick, and several other gentlemen, on seeing the books, mathematical instruments, etc., I had brought with me for my private enjoyment,

were so kind as to show me some attention. I had occasion to show them, in conversation, by the scientific publications of Europe, that I had been engaged in an extensive survey of Switzerland, which was interrupted by the revolution. Professor Patterson sent to President Jefferson an account of my former life, which I furnished at his request; and Mr. Clay, the Representative to Congress from Philadelphia, before setting off for Congress, in 1806, asked me if I should be willing to take a survey of the coast, to which I assented. (Letter published in the New York American, probably in February, 1827. Principal Documents Relating to the Survey of the Coast of the United States since 1816, published by F. R. Hassler, Superintendent of the Survey. New York: William Van Norton, printer, 1834. Octavo, pp. 1–180, I–III: folding map. Second volume of the Principal Documents Relating to the Survey of the Coast of the United States, from October, 1834, to November, 1835. Published by F. R. Hassler, Superintendent of the Survey. New York: William Van Norton, printer, 1835. Octavo, pp. 1–156, I–III (I).)

55. An interesting reminiscence of his career in this period is contained in the diary of John Quincy Adams for July, 1815, where there is described an interview by himself, with Mr. Gallatin, at that time United States minister in London, in which the latter spoke of Hassler, who had just left them.

"That is a man of very great merit. He was sent by the Government to Europe to procure the instruments for the general survey of our coast, but he has outrun his time and his funds, and his instruments cost eight hundred pounds sterling more than was appropriated for them; and he is embarrassed now about getting back to America. I have engaged Messrs. Baring to advance the money for the instruments, and he is to go for his own expenses upon his own credit. He has procured an excellent set of instruments." Adams's Memoirs, III, p. 248.

The circulars elicited by Hassler's plan are printed in the Transactions of the American Philosophical Society for 1812, II.

56. Report of Alexander Dallas Bache, Superintendent of the Coast Survey.

57. Adams's Life of Gallatin, pp. 349, 350. Henry Adams in this admirable biography has shown that Gallatin was one of Jefferson's strongest supporters in plans for the public enlightenment, and that he had an ambition of his own for the education of all citizens, without distinction of classes.

I had another favorite object in view [Gallatin writes], in which I have failed. My wish was to devote what may remain of life to the establishment, in this immense and fast-growing city [New York], of a general system of rational and practical education fitted for all and gratuitously opened to all. For it appeared to me impossible to preserve our democratic institutions and the right of universal suffrage unless we could raise the standard of *general* education and the mind of the laboring classes nearer to a level with those born under more favorable circumstances. I became accordingly the president of the council of a new university, originally established on the most liberal principles. But finding that the object was no longer the same, that a certain portion of the clergy had obtained the control, and that their object, though laudable, was special and quite distinct from mine, I resigned at the end of one year rather than to struggle, probably in vain, for what was nearly unattainable. Life of Gallatin, p. 648.

58. See Benjamin Franklin's Meteorological Imaginations and Conjectures, in the Memoirs of the Literary and Philosophical Society of Mansfield.

Communications made at Passy (France), in 1784, and reported in the Pennsylvania Packet (in Congressional Library) of July 18, 1786.

59. Life of Josiah Meigs, p. 27.

60. Maury's Life, p. 77.

61. Scott, Storm Warnings, London, 1883.

62. American Journal of Science, July, 1871.

63. Thirteenth Annual Report of the Secretary of the Smithsonian Institution, p. 34 (1858).

64. Twelfth Annual Report of the Secretary of the Smithsonian Institution, 1857, p. 26; also Twentieth Annual Report of the Secretary of the Smithsonian Institution, 1865, pp. 54–57.

65. 1883, History of the United States Signal Service, with catalogue of its exhibit as the International Fisheries Exhibition. London, 1883; Washington City, 1883; octavo, pp. 1–28.

66. John Quincy Adams, in his diary for November, 1825, describes an interview with his Cabinet, and the discussion which followed the reading of his message before it was finally revised for sending to Congress.

"Mr. Clay wished to have the recommendations of a National University . . .

struck out ... The University, Mr. Clay said, was entirely hopeless, and he thought there was something in the constitutional objection to it. ... I concurred entirely in the opinion that no projects absolutely impracticable ought to be recommended; but I would look to a practicability of a longer range than a simple session of Congress. General Washington had recommended the Military Academy more than ten years before it was obtained. The plant may come late, though the seed should be sown early. And I had not recommended a University—I had referred to Washington's recommendations, and observed they had not been carried into effect."

Such opinions as those of Mr. Clay were evidently very much at variance with those of John Quincy Adams and of his illustrious father, whose action in the constitutional convention of Massachusetts has already been referred to, and at variance as well, it would seem, with the opinion of the early Republicans, as with those of the Federalists. The views of Washington and Madison, as well as those of Jefferson and Barlow, on these subjects have already been referred to.

Mr. Adams, in commenting upon an address delivered by Edward Everett before the Columbian Institute, January 16, 1830, remarks:

I regretted to hear ... a seeming admission that the power of giving encouragement to literature and science was much greater at least in the State Governments than in that of the Union. Memoirs of John Quincy Adams, VIII, p. 171.

67. It is interesting to know that in 1827, Mr. James Courtenay, of Charleston, published a pamphlet, an urgent plea for the establishment of a naval observatory. I am indebted to Mr. William A. Courtenay for the opportunity to examine this rare tract, which has the following title:

1827. Courtenay, James. An | Inquiry | into | the Propriety | of | establishing | a | National Observatory. | = | By James Courtenay, | of Charleston, South Carolina | = |———| Charleston, Printed by W. Riley, 125 Church street |— | 1827. 8° pp. 1–24.

68. Proceedings of the American Association for the Advancement of Science, 1851, pp. 6, 48.

69. The idea of an Academy of Science with unlocalized membership and, like the Royal Society and the French Academy, holding advisory relations with

the General Government, appears to have been present in the minds of many of the early statesmen. Washington, in his project for a great national university, doubtless intended to include everything of this kind. Joel Barlow and Thomas Jefferson at the beginning of the century were engaged in correspondence "about learned societies, universities, and public instruction." John Adams in a letter to Cutler, dated Quincy, May 1, 1802, referred to a scheme for the establishment of a national academy of arts and sciences, in which Mitchill, of New York, was interested, and which was to come up for discussion at a meeting in that city in the following month. (Life of Manasseh Cutler, II, p. 87.)

70. See Appendix D, and also A. C. True's A Brief Account of the Experiment Station Movement in the United States, United States Department of Agriculture, Experiment Station Bulletin No. 1, 1889, pp. 73–78.

71. The following statements were made in a report of the committee of the House of Representatives, March 3, 1886:

The act appropriating script to the amount of 30,000 acres for each Senator and Representative in Congress for the endowment of colleges for the benefit of agriculture and the mechanic arts, which was passed in 1862, has been fruitful. Some of the States endowed single colleges while others divided the gift between two or three. There were 17,430,000 acres of script and land granted, and the fund arising from their sales is $7,545,405. This has been increased by gifts from the States and from benevolent individuals of grounds, buildings, and apparatus to the amount of $5,000,000 more. And the last reports show that these colleges employed more than 400 professors, and had under instruction more than 4,000 students. This donation of the public funds has been eminently profitable for the Government and the country. Many thousands of young men educated in science have already gone out from their colleges to engage in the practical duties of life, and the provision is made for sending out a continued succession of these for all future time. And as science is not limited by State boundaries, it makes but little difference for the common good which of three institutions or States these graduates come from; their attainments are for the common good.

72. See Appendix E, and also F. W. Blackmar's History of Federal and State Aid to Higher Education, etc., Washington, 1890.

73. The following is a list of those already in existence: State academies of science etc., 1890–

California.—The California Academy of Sciences, San Francisco, 1854.

Columbia.—The Affiliated Scientific Societies of Washington City; the Philosophical Society, 1871; the Anthropological Society, 1879; the Biological Society, 1880; the Chemical Society, 1889; the National Geographic Society, 1888.

Connecticut.—The Connecticut Academy of Arts and Sciences, 1799.

Indiana.—The Indiana Academy of Sciences, 1885.

Iowa.—The Iowa Academy of Sciences, Iowa City, 1875.

Kansas.—The Kansas Academy of Science, Topeka, 1868.

Maryland.—The Maryland Academy of Sciences, Baltimore, 1822.

Massachusetts.—The American Academy of Arts and Sciences, Boston, 1780.

Minnesota.—The Minnesota Academy of Natural Sciences, Minneapolis, 1873.

Missouri.—The St. Louis Academy of Science, St. Louis, 1857.

New York.—The New York Academy of Science, New York City, 1817.

Pennsylvania.—The American Philosophical Society, Philadelphia, 1743.

Wisconsin.—Wisconsin Academy of Arts, Science, and Letters, Madison, 1870.

74. The first agricultural experiment station under that specific designation in the United States was established at Middletown, Connecticut, in 1875, by the joint action of Mr. Orange Judd, the trustees of the university at Middletown, and the State legislature, with Professor W. O. Atwater as director, and was located in the Orange Judd Hall of Natural Science. The example was speedily followed elsewhere, so that in 1880 there were four, and in 1886 some seventeen of these institutions in fourteen States. The appropriation by Congress of $15,000 per annum to each of the States and Territories which have established agricultural colleges, or agricultural departments of colleges, has led to the establishment of new stations or the increased development of stations previously established under State authority, so that there are to-day forty-six stations in the United States. Several of these have substations working under

their management. Every State has at least one station, several have two, one has three, and Dakota has set the Territories an example by establishing one within her boundaries.

These forty-six stations employ nearly four hundred men in the prosecution of experimental inquiry. The appropriation by the United States Government for the current year, for them and for the Office of Experiment Stations in this Department, is $600,000. The several States appropriate about $125,000 in addition, making the sum total of about $725,000 given from public funds the present year for the support of agricultural experiment stations in the United States.

Of all the scientific enterprises which the Government has undertaken, [wrote Secretary Colman,] scarcely any other has impressed its value upon the people and their representatives in the State and national legislatures so speedily and so strongly as this. The rapid growth of an enterprise for elevating agriculture by the aid of science, its espousal by the United States Government, its development to its present dimensions in the short period of fourteen years, and, finally, the favor with which it is received by the public at large, are a striking illustration of the appreciation on the part of the American people of the wisdom and the usefulness of calling the highest science to the aid of the arts and industries of life.

75. The names of W. A. Chanvenet, J. H. C. Coffin, Mordecai Yarnall, Joseph Winlock, Simon Newcomb, Asaph Hall, William Harkness, and J. R. Eastman are a few of those to be found on this list of astronomers and mathematicians.

76. The secret history of this appointment is told as follows by Doctor Silas Reed, of Boston, in Lyon G. Tyler's Letters and Times of the Tylers (II, p. 696):

I called upon Mr. Tyler the next day, and found him about as well pleased over the result as I was, as it constituted a triumph that had never been achieved before (nor since), as shown by the annals of the Senate. While in this pleasant mood, the President asked me if I could not suggest some means by which he might soften the asperities of Senator Benton towards him and his administration. In an instant the thought flashed through my mind as to how he could best accomplish his wish. I said, "You have it in your power to touch his heart through his domestic affections. Six months ago his pride was humbled by the marriage of his highly educated daughter, Jessie, to a mere lieutenant of the

United States engineer corps, and he refused them his house. I have just learned that lately he invited them to return to his home, and know they have done so. Now you have a chance to gladden the senator's pride, and by so doing serve both yourself and the country, by taking Lieutenant Frémont by the hand, and giving him a chance to rise in the world by appointing him to head an expedition to explore the Rocky Mountains and some part of the Pacific coast."

Mr. Tyler thought it might stir an excitement with the higher grade officers of the engineer corps (as it did), and that he might not be fully competent to execute the high duties entrusted to him. I replied that these objections need not prevent his appointment, for Lieutenant Frémont had spent the last two years aiding the eminent French scientist, Nicolet in taking the hydrography of the valley of the Mississippi, and must be familiar with all instruments and modes of using them in such an expedition. And even if he should not prove judicious in selecting scientific men suitable for that part of his corps, he would have the able assistance of Colonel Benton and his talented wife to fall back upon; and that Senator Benton, on the return of Mr. Frémont, would receive, examine, and present his report to the Senate, and take great pride in making an eloquent speech of it (as he did), and thus cause the American reader to examine and well consider its instructive contents—all of which events took place, and the report of his first, if not his second, expedition gained sufficient notoriety to insure its republication in Germany.

At the close of our interview, the President, in his most earnest manner, said: "I will at once appoint Lieutenant Frémont to the head of such an expedition, and start him off this spring, so that the country may know as soon as possible what to say and believe of that vast and unknown region, and I shall learn how much effort to expend in striving to acquire it by purchase from Mexico by the time that Texas can be annexed."

Frémont made ready to start from St. Louis with his expedition as soon as there was green grass to subsist his animals upon, with an outfit of fifty to sixty men, after leaving Independence, Missouri, he moved up the Platte river and its north branches to the old "South Pass," and thence to the head waters of Snake (or Lewis) river, and down it and the Columbia river to Astoria, thus avoiding Mexican Territory, but kept close along its northern border until after he entered Oregon Territory.

77. Letters and Times of the Tylers, by Lyon Gardner Tyler, II, p. 387.

78. See J. S. Billings, Medical Museums, with Special Reference to the Army Medical Museum at Washington. President's address, delivered before the Congress of American Physicians and Surgeons, September 20, 1888.

79. See the eighteen annual reports of the Commissioner of Education.

80. See G. Brown Goode, The Status of the United States Fish Commission in 1884, etc., Washington, 1884.

81. See the six annual reports of the Bureau, and the Smithsonian reports, 1879–1888.

APPENDIX C

1. This is asserted in a book written to support the present government in France. I forget the title.

APPENDIX D

1. Introduced in the House of Representatives by the Hon. Justin S. Morrill, of Vermont, and approved by President Lincoln, July 2, 1862.

2. Introduced in the House of Representatives in 1885 by the Hon. William H. Hatch of Missouri, and approved by President Cleveland, March 2, 1887.

3. The grants of money to carry out the provisions of this act amounted in 1887–88 to $585,000, in 1888–1889 to 595,000, in 1889–90 to $600,000, and for 1890–91 the amount estimated is $630,000.

APPENDIX E

1. See Blackmar's Federal and State Aid to Higher Education.

2. State grants have been made to Bowdoin College, 1794–1802, and to Colby University, formerly Waterville College, 1818.

3. The appropriations by the State to Harvard have amounted to $784,793,

in addition to 46,000 acres of land. The State has also given $157,500 to Williams, and $52,500 to Amherst.—BLACKMAR.

MUSEUM-HISTORY AND MUSEUMS OF HISTORY

1. A paper read before the American Historical Association, in Washington City, December 26–28, 1888.

2. Owen, Transactions, Zoological Society of London, V, p. 266, footnote.

3. The collections of Sloane, who was one of the early scientific explorers of America, were like those of the Tradescants, contained many New World specimens, and the British Museum as well as the Ashmolean was built around a nucleus of American material. Indeed, we can not doubt that interest in American exploration had largely to do with the development of natural history museums.

In those days all Europe was anxious to hear of the wonders of the new-found continent, and to see the strange objects which explorers might be able to bring back with them, and monarchs sought eagerly to secure novelties in the shape of animals and plants.

Columbus was charged by Queen Isabella to collect birds, and it is recorded that he took back to Spain the skins of several kinds of animals. Even to this day may be seen in the old collegiate church in Siena a votive offering placed there nearly four centuries ago by the discoverer of America. It consists of the armor worn by him when he first stepped upon the soil of the New World and the rostrum of a swordfish killed on the American coast.

The state papers of Great Britain contain many entries of interest in this connection. King James I was an enthusiastic collector. December 15, 1609, Lord Southampton wrote to Lord Salisbury that he had told the King about Virginia squirrels brought into England which were said to fly. The King very earnestly asked if none were provided for him—whether Salisbury had none for him—and said he was sure Salisbury would get him one. The writer apologizes for troubling Lord Salisbury, "but," continued he, "you know so well how he [the King] is affected to such toys."

Charles I appears to have been equally curious in such matters. In 1637 he sent John Tradescant the younger to Virginia "to gather all rarities of flowers, plants, and shells."

In 1625 we find Tradescant writing to one Nicholas that it is the Duke of Buckingham's pleasure that he should deal with all merchants from all places, but especially from Virginia, Bermuda, Newfoundland, Guinea, the Amazons, and the East Indies, for all manner of rare beasts, fowls and birds, shells and shining stones, etc.

In the Domestic Correspondence of Charles I, in another place, July 1625, is a "Note of things desired from Guinea, for which letters are to be written to the merchants of the Guinea Company." Among other items referred to are "an elephant's head, with the teeth very large; a river horse's head; strange sorts of fowls; birds' and fishes' skins; great flying and sucking fishes; all sorts of serpents; dried fruits, shining stones, etc." Still farther on is a note of one Jeremy Blackman's charge—in all, £20—for transporting four deer from Virginia, including corn and a place made of wood for them to lie in.

4. This collection [we are told] was sold to Sir Ashton Lever, in whose apartments in London Mr. Adams saw it again, and felt a new regret at our imperfect knowledge of the productions of the three kingdoms of nature in our land. In France his visits to the museums and other establishments, with their inquiries of Academicians and other men of science and letters respecting this country, and their encomiums on the Philosophical Society of Philadelphia, suggested to him the idea of engaging his native State to do something in the same good but neglected cause.—Kirtland, Mem. American Academy of Sciences, Boston, I, xxii.

MUSEUMS OF THE FUTURE

1. A lecture delivered before the Brooklyn Institute, February 28, 1889.
2. Conway, Travels in South Kensington, p. 26.

SUBJECT INDEX AND GUIDE TO
FULL NAMES, INSTITUTIONS,
AND TITLES

This index is intended to make the historical studies of George Brown Goode more accessible by providing reference to individuals and topics. Only the text is indexed, but important themes in the appendices are also indicated. Over four hundred of the individuals in George Brown Goode's historical essays were mentioned only by surnames. Here full names and current preferred spelling are used for as many individuals as possible. Thanks are expressed for the thoroughness, persistence, and computer expertise of Horace Taft-Ferguson and the library skills of John T. Butler.